you will too
John
X

THE SECRET LIFE OF COWS

Further praise for *The Secret Life of Cows*:

'A wonderful book. Full of rare insights.' Sheila Dillon, presenter of BBC Radio 4's *Food Programme*

'Charmingly written. . . . whether you're a farmer or an urbanite, it's certainly a captivating read.' *Town & Country Magazine*

'Weaves entertaining anecdotes and profound insights harvested from a lifetime of caring for cows . . . A charming manifesto.' *Sunday Telegraph*

'This touching book will have you looking at Friesians in the field – and even a nice bit of rib-eye steak – completely differently.' *The Times* Books of the Year

'A little masterpiece.' *Oldie*

‘A plea for us to appreciate the complex inner lives of our inquisitive, loving, bovine friends, whom we arguably exploit more than any other creature on earth – from what we wear on our feet, via our Sunday roast, to what we pour on our granola. It also makes the great point that we should not judge animal intelligence in relation to our own.’ Matt Haig, *Guardian* Books of the Year

‘Couldn’t be more charming or compelling.’ *Herald*

‘This welcome reprint of a lovely little book is an antidote to the idea that domestic livestock are dull-witted, lower lifeforms . . . I certainly will look on cows with a new wonder from now on.’ *Countryfile*

‘The strength of Young’s cheering book is in the impassioned case she makes for farming animals with respect and kindness.’ *The Times*

‘A tender defence of animal intelligence.’ *i* newspaper

‘Humane, humorous . . . heart-warming.’ *National Geographic Traveller*

‘A meditative little book . . . Young’s style, careful and straightforward, is extremely soothing; her book should be prescribed for anxiety.’ *Observer*

‘A powerful pastoral meditation on animal husbandry, one for which Young should be congratulated and championed.’ *Sunday Times*

'As powerful an indictment of factory farming as anything I've read.' *Daily Mail*

'A beguiling book . . . Its power is in its honesty and genuine love of her animals. It's an argument for compassion and taking delight in the tiniest details of the natural world – observed only through quiet, persistent observation – that is charmingly persuasive.' *Independent*

'On Rosamund Young's farm, the cows live well. The result? They reveal their true and lovely nature. A fascinating insight into how we all respond better when nicely looked after.' *Sainsbury's Book Club*

'In this conversational book, she reveals the delightful foibles and frolics of her herd [and] militates for the compassionate care of these sociable, family-oriented creatures.' *Sunday Express*

'Delightful . . . Young has created a James Herriot-esque profile of her cows – all with distinct personalities and idiosyncrasies . . . Young's book is a reminder that these animals deserve our care, respect and understanding.' *Emerald Street*

'Every farm should be like this. The animals have space and liberty.' Jane Grigson, *Observer*

'Absorbing, moving, and compulsively readable.' Lydia Davis

'Her insight is unexpectedly charming and fascinating – so much so that she has channelled everything she has learned about these creatures into a wonderfully evocative and enlightening book . . . Many amusing anecdotes which illustrate the warmth of feeling between the family and their individual animals . . . A wonderful and heartwarming story.' *Mail on Sunday*

The Secret Life of Cows

ROSAMUND YOUNG

FABER & FABER

First published in 2003 by Farming Books & Videos Ltd

First published in 2017
by Faber & Faber Limited
Bloomsbury House
74–77 Great Russell Street
London WC1B 3DA
This paperback edition first published in 2018

Typeset by Faber & Faber Limited
Printed in the UK by CPI Group (UK) Ltd, Croydon, CR0 4YY

A CIP record for this book
is available from the British Library

ISBN 978–0–571–34579–3

10 9

Some of my first memories are of one or other of my parents relating 'stories' that had occurred involving cows or pigs or hens or wild birds. I hope that I am continuing here what began as an oral tradition.

ROSAMUND YOUNG, KITE'S NEST FARM

AUTHOR NOTE: While putting this book together I was reminded that books have chapters. However, most of my anecdotes weave in and out of one another to form a narrative, making individual chapters unnecessary and cumbersome. Instead, headers between sections guide the reader through the text. This is a reissued edition of my book so I have had the opportunity to update some of the material. R.Y.

Contents

'Nature teaches beasts to know their friends.'

SHAKESPEARE: *Coriolanus*, Act II, scene i

'People watch with amazement a television programme on the social lives of elephants – their family groupings, affections and mutual help, their sense of fun – without realising that our own domestic cattle develop very similar lifestyles if given the opportunity.'

JOANNE BOWER, *The Farm and Food Society*

Foreword

When I came across *The Secret Life of Cows* I thought the title was a joke. But it isn't and it's about exactly that. It's a delightful book, though insofar as it reveals that cows (and indeed sheep and even hens) have far more awareness and know-how than they have ever been given credit for, it could also be thought deeply depressing, as it means entirely revising one's view of the world.

Had the book been written simply by an enthusiast one could dismiss it as the work of a crackpot but Rosamund Young has been running her organic farm at Kite's Nest in Worcestershire since before organics started. It's a farm where the farm hands can tell from the taste alone which cow the milk comes from, and Young makes the case against factory farming more simply and compellingly than anyone I've read and wholly on grounds of common sense.

One curiosity about the book is that while the author goes into much detail about the behaviour of cows and their differences of temperament and outlook she never mentions any idiosyncrasies about when the cows go with the bull and whether their individuality, which she has made much of elsewhere, is still in evidence. Are some shyer than others? More flirty? It may be that her reticence in this regard is a measure of her respect for her charges with the feeling that cows are entitled to their privacy as much as their keepers.

Still it's a book that alters the way one looks at the world, with dumb animals not as dumb as we would sometimes like to think. It's a book that alters the way one sees things and passing a field of cows nowadays I find myself wondering about their friendships and their outlook, notions that before reading Young's book I would have thought fanciful, even daft. Not any more.

ALAN BENNETT

Introduction

Watching cows and calves playing, grooming one another or being assertive, takes on a whole new dimension if you know that those taking part are siblings, cousins, friends or sworn enemies. If you know animals as individuals you notice how often older brothers are kind to younger ones, how sisters seek or avoid each other's company, and which families always get together at night to sleep and which never do so.

Cows are as varied as people. They can be highly intelligent or slow to understand; friendly, considerate, aggressive, docile, inventive, dull, proud or shy. All these characteristics are present in a large enough herd and for many years we have been steadfast in our determination to treat our animals as individuals.

My mother and father started farming in their own right in 1953. My brother Richard was nearly three and I was twelve days old. To begin with they had five cows and an old tractor, no electricity and no telephone.

They gradually built up a herd of pedigree Ayrshires and also kept Wessex Saddleback pigs. There were enormous numbers of rabbits on the land, which made crop-growing impossible.

Financial incentives in those days depended on intensification; there was undisguised pressure from government departments for farmers to use every modern aid. My parents' instincts were to be organic, although they had never heard the word, and the process of going against the official line happened gradually. From the start they both shared an absolute determination to make the lives of the animals in their care dignified and comfortable.

Some of my first memories are of one or other of my parents relating 'stories' that had occurred involving cows or pigs or hens or wild birds. I hope that I am continuing here what began as an oral tradition.

Cows are individuals, as are sheep, pigs and hens, and, I dare say, all the creatures on the planet however unnoticed, unstudied or unsung. Certainly, few would dispute that this is true of cats and dogs and horses. When we have had occasion to treat a farm animal as a pet, because of illness, accident or bereavement, it has exhibited great intelligence, a huge capacity for affection and an ability to fit in with an unusual routine. Perhaps everything boils down to the amount of time spent with any one animal – and perhaps that is true of humans too.

Everyone who keeps just a few animals will unquestionably know them as individuals and will probably talk about the finer points or idiosyncrasies of their natures with much understanding. Farmed animals are usually kept in large groups but this does not mean that individuality disappears. Their levels of intelligence vary just as much as is true of human beings.

No teacher would ever expect or want all the pupils in one class to be identical. No one would want to create a society in which everyone wore the same clothes or had the same hobbies. Just because we are not clever enough to notice the differences between individual spiders or butterflies, yellowhammers or cows is not a reason for presuming that there are none.

Animals and people can appear to lose their identities or become institutionalised if forced to live in unnatural, crowded, featureless, regimented or boring conditions. When this happens, it is not proof that individuals are all the same or want to be treated as such.

Many people judge the comparative intelligence of different species by human standards. Yet why should human criteria have any relevance to other species? We should presume that every animal has a limitless ability to experience a whole range of emotions, judged only on its own terms. If a cow's intelligence is sufficient to make her a success as a cow, what more could be desired?

If, when a young calf tries to eat some hay, it is repeatedly pushed away by bigger, stronger cattle and it then works out that by squeezing in under its mother's chin it will be able to eat in peace, that seems to me an example of useful practical intelligence. What would be achieved by teaching the same calf to open a gate by pressing a button with its nose? Nothing.

During a lifetime observing cattle I have witnessed amazing examples of logical, practical intelligence and some cases of outright stupidity, both of which qualities I have also remarked in respect of human beings. Cattle merely get on with the day-to-day business of living, solving or failing to solve problems as they arise. The important point is that they should be given the wherewithal to succeed as animals, not as some inadequate servants of human beings.

The assertion I once noted in *Star and Furrow* magazine, that restricting a cow's ability to move freely would, after a few generations, result in a 30 per cent reduction in brain size, ties in interestingly with our own observations. In the 1970s my parents noticed that their cows' foreheads were getting wider and that the cows looked and indeed behaved more intelligently. Some ten or fifteen years later, and quite by chance, we were visited by a scientist who worked for one of the country's largest zoos. He was exclusively occupied, and had been for the twenty years preceding the first acknowledged case of BSE, in examining and specifically measuring the craniums of dead animals. He had chronicled a relentless shrinking of brain size during this period and had concluded that it was caused entirely by what he described as the atrocious and probably BSE-infected food on which the animals were fed. I now consider it likely that the incarceration of the animals could be equally, if not more, to blame.

Meat quality also is affected by diet and freedom. There are higher levels of omega-3 polyunsaturated oils and a lower fat-to-protein ratio in meat from animals that enjoyed a wild diet in comparison to those reared in an intensive manner.

No one would expect a child to develop normally when kept in cramped, unfriendly conditions, deprived of parents and siblings and with restricted exercise and the same diet every day, yet many farmers and the government departments that inform them seem to expect farm animals to develop normally in such circumstances.

For many years we have noticed that if you give cows the opportunity and the time to choose between several alternatives – for instance between staying outside or coming in for shelter, or walking on grass or on straw or concrete, or a choice of diet – then they will choose what is best for them and they will not all choose the same thing.

A hen loves to run and investigate everything that moves, to spread its wings in the sun and preen its feathers and bathe in the dust. It must never be confined in a tiny cage or overcrowded building. The claim that some 'free-range' hens do not choose to venture outside even when the portholes are left open for them is quite invalid once it is realised that there isn't always enough grass outside to make it worthwhile and that in large flocks hens at the bottom of the pecking order can become very intimidated and frightened to venture outside.

The mutilation of farm animals has its roots in propaganda, custom and thoughtless adherence to tradition and it cannot be justified on any grounds. The routine tail-docking and teeth-cutting of piglets, the debeaking of poultry and the tail-docking of sheep is indefensible.

If pigs bite or hens peck one another it is because they are unhappy, and if lambs' tails get dirty – I have seen at first hand the misery of dealing with maggot-infested sheep – the cause needs to be addressed; it is not a solution to cut off the tail.

Making animals happy and allowing them to express their natural behavioural instincts is not just morally and ethically essential; it also makes sound financial sense. Happy animals grow faster.

Children under stress eat and sleep less well than those who are happy and relaxed. Unhappy children develop real and imaginary ailments such as headaches, eczema and weight problems. Stress can be reduced or eliminated by improving existing conditions. A change of environment or diet and more understanding or love all play their part; it is the same with animals.

It is misplaced conceit to believe that any man-made environment can equal or better the natural one. Piglets are often weaned when they are far too young and transferred to seemingly warm and safe accommodation. No artificially manufactured conditions can match the reassurance, stability, attention, companionship and appropriate food that nature provides. As a result, it is at this point that they frequently become ill and are given their first course of antibiotics.

Success in farming is increasingly measured in terms of output. High output figures are recorded and success is assumed if a female animal produces a large number of offspring within a short space of time. However, what is not taken into account is the fact that the almost constantly pregnant mother might well have a reduced life-span and will not have the opportunity to pass on to her progeny her own accumulated wisdom because of unnatural, forced weaning strategies. This increases the chances that future generations will be less knowledgeable and less well equipped to deal with maturity or motherhood themselves. This is farming for the short term.

Einstein said that 'the only really valuable thing is intuition'. Instinct and intuition are the most useful tools any living creature possesses. Yet in virtually all intensive farm livestock enterprises they are ruthlessly suppressed and all possibility of their developing is blocked. We suppress instinct in animals and children at a huge risk to the whole community.

Wherever the pursuit of maximum profit has led to

intensification it is the animals that have suffered most. Livestock diseases are often caused or exacerbated by overcrowding, inadequate housing and poor or dangerous feed quality. The living conditions created within these systems cause stress, and it is widely recognised that the production of stress hormones reduces the efficiency of the immune system.

Where cattle have adequate living space, freedom from competition for food, licence to roam freely and, above all, where they can live in family groups in which there is a preponderance of mature animals, immunity to lung and stomach worms can be established. This obviates the need for anthelmintics, which compromise the natural ability to resist such parasitic infections and can leave residues in meat and milk.

On farms where animals are grouped by age or size, they are deprived not only of health benefits, but also of the company of older animals, from whom, in a more natural environment, they would learn. Many cows live totally unnatural lives, and dairy cows in particular are much misused animals. Frequently regarded solely as providers of milk, most dairy cows are fed in such a way as to maximise their output. Often the high-protein diet takes no account of the cows' preferences, physical or dietary requirements, comfort or long-term health. Immediately after being born, calves are usually forcibly taken from their mothers and reared in a variety of unnatural ways, or simply shot. They are often fed on a milk substitute, rather than the cows' milk that is their birthright. They are frequently housed unsuitably in tiny, individual pens or kennels where they have no contact with others of their species. Some pens are permanently under cover, depriving the calves of fresh air, sunshine and exercise. Feeding regimes prevent calves from eating or drinking as and when they need to.

Lameness, which is often chronic, affects a large proportion of dairy cows and much of their life is spent walking and standing on unsuitable and uncomfortable surfaces. When cows are in pain they eat less, if at all, and frequently become infertile, as a result of which they are culled from the herd.

Many cows are kept permanently inside, often in very large numbers in so-called mega-dairies. Milking cows in these systems never graze grass, see a field or leave their area of confinement. The quality of the resultant milk is questionable; the quality of the life of the cow is at best unnatural and at worst unbearable.

At our farm, Kite's Nest, the calves stay with their mothers for as long as they choose. They suckle milk for at least nine months, effectively weaning themselves when the cows' milk dries up, between one and three months before the next calf is due. Back in 1953 we milked a commercial herd of pedigree Ayrshires but by 1974, we realised that not only was milk production uneconomic (and very hard work) but my parents, my brother and I were seriously questioning whether such a system was truly the one we felt happy with. We decided to change to a single-suckle system where each cow rears her own calf.

The very significant turning point came in the early 1980s on the day an especially clever and delightful young bullock called Lochinvar was due to leave the farm to be sold as a store, for someone else to fatten for beef, in the local livestock market. We all knew just how much we would miss him and we couldn't help dwelling on what his subsequent home might be like: whether he would have to travel too far and become hungry or thirsty: whether he would be treated kindly. That was the day we decided to retail beef from our own farm shop so that we could be in charge of every stage of production and be able to give our customers the assurance that we knew precisely what the animals had been fed on for the whole of their lives. As recently as 2012, we started a commercial flock of sheep and now also sell lamb and mutton.

When it comes to milk most calves will know where to look for milk and how to suckle but on occasions at the very beginning they might need help from us. If parturition has been trouble-free and relatively painless, calves will suckle from a 'normal' position, whereas if the calving has been difficult and painful the cow will sometimes appear to hold the new arrival responsible for the pain and permit it to suckle only from behind, out of sight.

Calves play games together, copying and learning constantly. They learn where the best and sweetest water is to be found and how to nibble young shoots on the hedgerows. Like cats and dogs and people, and I suppose every other creature, calves learn whom they can trust. A calf will settle itself down under the nose of an older animal only if it knows it will not be bullied. Some bovines are bossy; they seem to need to maintain an air of dominance, even if only by an unprovoked and often fairly gentle push, while others are always good-tempered and some are quite timid. Everything in a herd of cows comes down to character: a polled cow can often deter a horned one with a mere look.

Cows and calves go about their daily lives in as many different ways as human mothers and children do. Some form such close friendships with each other that the calves do not branch out on their own for weeks. Many make firm friends with other tiny calves at only one or two days old. Generally these attachments complement the affection they have for their mothers, but they have been known virtually to replace it, as in the case of 'the White Boys' (see page 59), who acknowledged their dams only when they wanted milk or to be groomed.

All the cows and calves in these anecdotes have had a great deal of freedom and have never been prevented from choosing whether they want to be outside or inside. They have also had permanent access to food and water.

All unnecessary confinement and unnatural treatment of animals is indefensible. Farm animals that are allowed sufficient freedom can choose healing plants for themselves and stress-free routines that obviate the need for any routine medication. Our animals do indeed seek out the plants they sense they need. Bovines regularly go blackberry-picking in the autumn and eat young hawthorn leaves and shoots in the spring; they eat ash leaves and willow whenever they get the chance. Some will search for wild thyme and sorrel while others, at particular times of year and often depending on their stage of pregnancy, will eat large quantities of stinging nettles. Sheep will eat thistles and dock leaves voraciously by choice. In *The Shepherd's Calendar*, John Clare wrote: 'The ass . . . will eager stoop/ to pick the sprouting thistle up.' Docks are deep-rooted and their leaves contain important minerals and other trace elements not readily available from shallow-rooting plants.

We decided that the animals themselves are by far the most qualified individuals to make decisions about their own welfare and it is the decisions they make, as well as many other occurrences both humdrum and extraordinary, that I have observed, learned from and written down here.

Farmers have a clear moral obligation towards their animals but it is interesting to note that the meat from animals reared extensively actually tastes better and is held by many doctors and other individuals to be better for you.

When Boswell wrote his *Life of Dr Johnson*, he asked 'whether such a mind as his was not more enriched by roaming at large in the fields of literature, than if it had been confined to a single spot', and drew the following analogy: 'The flesh of animals who have fed excursively is allowed to have a higher flavour than that of those who are cooped up.'

In Thackeray's *Vanity Fair*, the character Lord Steyne, having recently dined with the king on a joint of mutton, says, 'A dinner of herbs is often better than a stall-fed ox.'

In more modern times, Dr Peter Mansfield has assessed the role that fat plays in our diet and whether it is responsible for the increase in heart disease. In *Chemical Children*, he writes:

More plausible is the suggestion that the kind rather than the amount of fat is wrong . . . We have already learned a lot by reflecting on the habits of our ancestors, many of whom thrived well on meat. But their animals were healthy and enjoyed a wide variety of food in all seasons. They ate grass not only young and green, but in seed a few months later. And they browsed leaves and shoots off trees as far up as they could reach. They still do, when they get the chance. But trees grow slowly, and . . . pastures are usually fenced-off grass enclosures nowadays. No grass gets a chance to seed . . . So modern animals enjoy a much narrower variety of food than their ancestors . . . Whole seeds and dark green leaves are conspicuous absentees. In time past these would have been their principal sources of two special fatty acids which they cannot make from any other source: linoleic acid and linolenic acid . . . Without these the animals do not grow . . . The fat composition of modern diets is clearly quite important and may prove relevant to a whole range of conditions which puzzle us now, from allergy to multiple sclerosis . . . Animals not bred for health prove at times not healthy enough . . . We should be exposed to much less immediate or potential hazard if it [meat] were grown slowly, for health rather than quantity.

The difference between intensive and organic systems of rearing and caring for hens is enormous and the gap is widening daily. For example, a bird destined to become a Sunday roast will reach its target weight in approximately 80 days in an organic system and only 42 in an intensive one. The intensively reared birds are frequently given antibiotics, to help prevent them dying from the effects of the overcrowded and unnatural conditions in which they live. They are denied fresh air and daylight, packed so tightly that exercise is virtually impossible, while the way in which breeding programmes have prioritised growth rates and size means their skeletons are often unable to support their own weight. This results in a high incidence of broken bones. Too heavy and too deformed to perch, the birds spend their lives in ammonia-soaked dung, which burns their hocks and feet.

Farmed birds should be allowed to follow their behavioural instincts, grow at their natural rate, eat safe food and live dignified lives. Almost all the pullets destined for egg production are hatched in incubators and reared in artificially heated pens. More than half of them are still transferred at 'point of lay' to wire-mesh cages. Here they live out their days unable to fulfil any natural function or instinct, having first had their beaks cut to deter them from pecking their cellmates out of boredom.

A hen in a natural situation would lay her eggs and sit on them until they hatched. She then would teach her chicks what to eat and how to find the food. She would guard, befriend and protect them, constantly talking to them and constantly on the look-out for danger, which, once sensed, would prompt her to round them up and hide them in double-quick time. And research shows that chicks reared without a hen are more aggressive.*

When hens are healthy and happy, their feathers shine, their eyes are bright and alert, and they spend all day being as busy as bees: pecking, grazing, chopping, pull-

* Falt, B. (1978) 'Differences in aggressiveness between brooded and non-brooded domestic chicks', *Applied Animal Ethology* 4, 211–21.

ing, running, investigating, digging, playing and singing contentedly. If they cannot find what they need and there are humans handy they know and trust, they will come close and sing ever louder until they are noticed. Then it is up to the human being to try to work out what is wanted. This is not usually very difficult.

If hens are unhappy and unhealthy their feathers are dull; they do not sing; they moult too often. They stand hunched up and can become over-timid or over-aggressive. Health is established and maintained by giving hens what they need and poor health is invited by deprivation. It appears that when hens and pigs are confined in small spaces and given no choice about what they eat, they will eat what they are given and carry on living, however bored and frustrated they might be. If a cow or sheep were confined in a commensurately small space, I dread to think how she would cope – or indeed if she would cope at all.

It is widely accepted that animals such as cats, dogs and horses, usually kept in small numbers and given individual attention, are capable of exhibiting symptoms of boredom and unhappiness, that they can pine and grieve and show signs of feeling unwell. Hens are usually kept in such enormous flocks that monitoring individuals is impossible and it is generally considered that because they cannot attract the attention of their keeper they do not have feelings that matter.

It is perhaps easier to assume that animals have no feelings. They can then be used as generators of profit without any regard being given to their actual needs, as satisfying those needs is allegedly not worth the cost. Happy animals grow faster, stay healthier, cause fewer problems and provide more profit in the long run, when all factors, such as the effects on human health and the environment are taken into account. W. H. Hudson said, 'Bear in mind . . . that . . . animals are only unhappy when made so by man.'

Bovine needs are in many respects the same as human ones: freedom from stress, adequate shelter, pure food and water, liberty to exercise, to wander about, to go for a walk or just to stand and stare. Every animal needs congenial company of its own species and a cow needs to be allowed to enjoy her 'rights' in her own way, in her own time and not according to a human timetable.

The number of different ways a calf may be treated is no fewer than the number of ways a child may be treated. Most people believe that children need a stable environment with warmth and comfort, good clothes and shoes, food and drink, interesting diversions, friends of their own age and adults to guide and, above all, to love them. We do not expect a well-balanced adult to emerge from a neglected, ill-nourished, lonely, frightened child. I believe we should apply the same logic to farm animals. The quality of the food and the overall environment of any living creature will determine its potential in later life.

Pigs are often described as intelligent and indeed they are. It is iniquitous criminality to keep them in confinement, preventing them from making nests and rooting in the earth. Intensive systems change them from being good-tempered, happy creatures into angry, dangerous, disturbed but still intelligent individuals.

Sheep are equally often described as stupid or silly, which they most certainly are not, as George Henderson so astutely observed in *The Farming Ladder*: 'Contrary to the common opinion the sheep is by far the most intelligent of all the farm animals.' I was once given an orphan lamb and I named her Ellen. She was brought here when just two hours old, having been given colostrum from her mother, who did not have sufficient milk to rear both her two lambs. The farmer who brought her has a distinctively deep, rough voice. Six weeks later he called again and Ellen recognised his voice and ran to him. Several years later, when I banged my knee and was hopping about in pain, she left her food and came over to me with evident compassion and would resume eating only once I had succeeded in convincing her I was no longer in pain, even though I was!

Various factors have conspired to make the keeping of large flocks of sheep essential from a financial point of view. When hundreds of sheep are all kept together it is next to impossible to identify any individual attributes, but if you make enough time to observe sheep as individuals you will see their personalities emerge and develop.

The behaviour and health of all animals is affected by the quality of food they receive and the stress to which they are subjected. Stress can be caused by the food itself. However innocuous any one chemical applied to the growing crop, or any single additive applied to the end product, foods are usually eaten in combination and the effects on health of consuming several unnatural substances together are largely unknown. With the vast majority of intensively reared animals in the UK now being fed on genetically modified soya-bean meal, only time will tell if this results in additional problems.

When there is competition for food and the weakest get insufficient they will be under additional stress, which increases the risk of disease developing. In large flocks it is very difficult to treat ill sheep individually, so mass medication is widely used, often preventatively. While this can make life easier for the shepherd, it generally increases the overall amount of medication each animal absorbs, which is one of the factors behind rising levels of resistance to some drugs.*

* Official Journal of the European Union, Commission Notice Guidelines for the prudent use of antimicrobials in veterinary medicine (2015/C 299/04).

The incidence of tuberculosis in cattle, considered by some to be caused by badgers, is an issue that comes to mind when dealing with the subject of stress, and which was noted in evidence to the House of Commons Agriculture Committee inquiry into badgers and bovine tuberculosis in 1999: 'Cattle kept in intensively managed herds are exposed to conditions comparable to those which encourage the spread of human TB in the poorest and most overcrowded parts of the world.'

We have tried on this farm to create an environment that allows all of the animals the freedom to communicate with or dissociate themselves from us as they choose.

The seemingly mundane, day-to-day existence of a cow or a calf or of both together is perhaps not a subject that would capture everyone's imagination. These true stories give the reader a glimpse of what goes on in the lives of ordinary bovines as they pursue their daily routine, and reveal these lives to be as full and varied as our own. Although much of the day has to be spent working (i.e. eating) to sustain themselves, time can always be found for extra-curricular activities such as babysitting for a friend, blackberry-picking, fighting a tree or bank of earth, playing tag with a group of youngsters or a fox or quietly discussing an impending confinement with a daughter. All these activities and many more have been observed by us over the years and this selection of stories is a record of a hitherto secret world.

I have told the stories exactly as they happened but of course the interpretation of the actions of the 'characters' is mine. I have deliberately used personal pronouns when talking about the cows because that is how I think of them.

The Secret Life of Cows

We had always been proud of our herd of cows. We milked them, spoke to them by name, stroked them and generally enjoyed their individuality. But I was fully thirteen before I realised that they liked each other.

In 1968 we were dairy farming with a herd of pedigree Ayrshires. That summer we rented three fields on a steep, unspoiled hill four miles away and hired a lorry to take the dry cows and heifers to their summer grazing. They stayed there for three months, eating lush grass, drinking ice-cold spring water and altogether enjoying themselves. The left-at-home milkers seemed as happy as usual too. When the expiry date on the short-term tenancy on the hill drew near, we booked the same lorry and on the appointed day we brought the holidaymakers home again.

I believe all four of us noticed that for several days after the two halves of the herd had been reunited, Sunbeam and Moonbeam, mother and daughter, stood shoulder to shoulder in the yard and in the field talking over the last three months, not exhibiting any emotion but just very glad to see each other again.

They had not pined when they were parted. As a milking cow, Sunbeam had not reared her daughter and we did not even know that they recognised each other, but that demonstration of mutual affection opened our eyes to a whole new world, the world of bovine friendships.

A little bit about ingenuity

When Wizzie, also an Ayrshire, had her second calf (a chunky, pretty, short-legged strawberry roan heifer called Meg), she told her daughter she was the best and the calf believed her. Once winter set in and mud became an everyday problem, Meg made it clear that she hated getting her mahogany-coloured feet dirty. Somehow she managed to negotiate the steep flight of a dozen narrow, Cotswold-stone steps up to the granary and early one frosty-cold morning we watched her come out onto the top step, yawn and look around to see if it was worth getting up – i.e. coming down. She had spent the night in great comfort on the wooden granary floor, away from mud and draughts and bullying. We had left the granary door open because we *knew* that no bovine could possibly climb the steps. Subsequently she taught two friends the same trick and we used to put hay and water upstairs for them.

Alice and Jim

We stopped milking cows commercially in 1974, from then on allowing the cows to rear their own calves, but we still milked one or two cows for our own domestic consumption.

Alice became one of our house cows in 1990 and during the time we spent walking her in every day and milking her we discovered not just how intelligent, utterly kind and gentle she was but also what a sense of fun she possessed.

Alice was big and black with a wide, intelligent forehead and large, dark eyes, and she was quick to learn the milking routine. We milked only once a day, our aim being self-sufficiency rather than quantity. Every day, in the early evening, one of us would walk to fetch the milkers. They were nearly always in their favourite L-shaped field. This field commands one of the best views on the farm, is flatter than any other field and seems to go on for ever, but we are not sure whether the view has anything to do with the preference. From the farmyard you have to walk uphill through the Walnut Tree Field, home to five 120-year-old trees, and when you reach the top, the L-shaped field stretches out before you. More often than not both the house cows would be as far away as they could possibly be. But they knew why we had come and would walk back to the farm quite happily.

Sometimes Alice would liven things up a bit and from ambling by my side she would suddenly change speed, kick up her heels and disappear out of sight. I would continue to amble with the other house cow and then several hundred yards further on would spot Alice trying to play hide and seek. She would do her best to hide behind a walnut tree but of course she was too big and as soon as she realised I had seen her she would gallop off again and hide behind the next one, and so on until we reached the cow pen.

After her year as a house cow, Alice had a three-month rest out in the fields with her friends. As it drew near the time for her to calve again, we walked her down to the barn so that we could be close at hand in case she needed help. Alice realised we had come to take her home and seemed happy to comply. However, after fifty yards or so at walking pace she suddenly accelerated and dashed off to the other side of the field. She flew over to her friend Toria and told her where she was going and why, then trotted back to where she had left us. We finished our journey home uneventfully, and the next morning Abou was born, without our assistance. Toria had Gloria a week later and all four were soon reunited with the herd.

The following year Alice had Jim. He was jet black with a white tail and a very high IQ. When Jim was a year old, his twin siblings, Alice II and Arthur, were born. By this time Jim had weaned himself physically and emotionally from his mother and had taken himself off into another part of the farm with his friends. He kept half an eye on his mother and when Alice brought the twins out into the sunshine for the first time and left them for a while to go and graze, Jim jumped the fence and half trotted, half marched over to introduce himself to them. They were much too small to be of any real interest to him so he turned and trotted back. Arthur, barely twelve hours old, decided to follow. His legs were still a bit unsteady but his determination not to be left behind grew with each step and we watched as he bumbled and stumbled and hurried after his big brother. Jim reached the fence, popped over and was gone. Arthur stared in unbelieving disappointment, examined the fence from every angle then slowly made his way back to his sister.

A few months later, in gloomy winter, Jim worked out a way to make each day far more interesting than it would otherwise have been. Fat Hat II was living in the same barn as Jim and, as usual, was allowed to go out whenever she asked, always choosing to head for the silage face and eat ad lib for an hour or so. Jim could not understand why he was not offered this privilege too. After watching carefully for a few days he worked out a solution that amused and amazed us all.

I was standing in the kitchen with a friend, probably drinking tea and certainly looking out of the window, when we saw Jim walk out of the yard into the field. The gate had been left open, but as it was so cold none of the others had ventured out. He walked single-mindedly away from his friends and food towards the Cherry Tree Field. It suddenly occurred to me what he was going to do, and I gave my friend a running commentary as Jim continued walking for about a hundred yards, turned 180 degrees, tiptoed over the cattle grid, walked along the road in front of the house and joined Fat Hat II at the silage clamp. Needless to say, from that day he too was allowed to go by the shorter route.

Mothers and daughters

Relationships between mothers and calves are often complicated and fascinating. Some mothers are mild and bossed about by their calves; some are overbearing; others too casual. But perhaps two of the more interesting stories concerned Dolly and Dolly II and Stephanie and Olivia.

Stephanie and her daughter Olivia enjoyed a normal, close relationship and went everywhere together until Olivia had her first calf. When the calf was due to be born, Stephanie advised and comforted Olivia and helped her choose a good spot to calve, close to clear, running water. Stephanie settled herself down at a handy but not intrusive fifty-yard distance. Olivia calved without difficulty and was immediately besotted by her beautiful cream-coloured bull calf, whom we named Orlando.

She licked him dry, suckled him and quite simply doted on him. Stephanie came along a couple of hours later to be introduced and for the next few days grazed nearby hoping to be a useful and integral part of the threesome. As young calves spend a great deal of time sleeping in the first few days, grandmothers are often useful for babysitting. Sometimes cows who are not related are called on to babysit. It is quite common for one cow to look after several calves at once, but the job allocation is done democratically and cows take it in turns.

Sadly, Olivia did not want Stephanie's services. She did not wish to stir from Orlando's side. She ate as close to him as possible and whenever he moved she followed. She even refused her mother's offer of grooming. She ignored her shamefully. On the fourth day Stephanie's patience broke. Hurt and amazed, she turned tail, jumped the nearest fence and went off into another field to graze with her erstwhile friends.

To the best of my knowledge they never spoke to each other again.

The case of Dolly and her daughter was altogether different. Dolly was a wise, fairly old cow. She was dark mahogany, slim, neat and very, very clever. She had had many calves and had looked after each one superbly. She gave them four or five gallons of milk a day for several months, gradually reducing the amount over a nine- to twelve-month period so that when the time came for them to be weaned they were deriving their basic diet from grass and hardly missed the milk. She groomed every inch every day. She protected and encouraged, and told them all to be wary of human beings. 'They are not like us,' she told them. 'They have their uses, occasionally, particularly for carrying hay in the winter, but there is absolutely no obligation to fraternise.' They all heeded this advice.

Her first four calves were boys and they lived in magnificent isolation from or, more accurately, indifference to us. Dolly's fifth calf was a girl, Dolly II.

Dolly II was very beautiful. She was as pale a gold-brown as her brothers had been dark. She had big, deer-like eyes and a sweet and trusting nature. No matter what old Dolly said or did, young Dolly liked us and liked us to like her. Sometimes we felt encouraged to give old Dolly a pat when we were stroking her calf, a sort of pat of congratulation. She would toss her head angrily as if we had forgotten the rules. Although we were pleased to be trusted by most of our cows, we admired the few who were independent from us.

When Dolly II was fifteen months old her mother had another calf and, true to previous form, devoted herself to it. Dolly II was not spurned but was increasingly ignored until she understood that as an adult she must make her own friends and leave her mother to the job she was so good at. She found it easy to make friends.

When Dolly II was getting close to having her first calf, we looked at her every day and as the time got closer we went twice and then three times a day. We always try to be on hand in case we are needed, although we seldom are. Each time she greeted us with friendly unconcern. She felt fine and could not understand our frequent visits.

We were not there when Dolly II calved. So far, nothing had ever gone wrong in her life and she did not expect it to. Instead of choosing an open, accessible place in which to calve, or walking home to ask us for help, as several young cows had done before, she went as far away from home as she possibly could and settled down, hidden from all sides by hedges and hills.

When we discovered she had disappeared, we knew why and began searching everywhere. This is quite a big farm and there are endless hiding places. If you are unlucky enough to look in all the wrong ones first it can take ages to find what you are looking for. There were five of us looking on that day and we all went in different directions with very specific orders.

Dolly II was finally found behind the hill in the Monument Field and she was a sad sight. The alarming truth was that in making a huge effort to produce, unaided, a much-too-big bull calf, Dolly had displaced her womb. The calf had been born dead, and when we found her, she was lying down, exhausted. We set to work to try to make her more comfortable. While waiting for the vet to arrive we gave her a drink of water with the chill taken off (less of a shock to her system than cold water) and covered her with a blanket. The vet arrived quickly and managed to reposition the womb and stitch it in place. We then propped her up into a sitting position with bales of hay and straw and finally left her looking relatively comfortable but still tired and seemingly unable to stand.

When we went back to see her an hour later the blanket was in a heap on the grass, the bucket was empty and tipped over and Dolly II was nowhere to be seen. We could not believe our eyes.

After much searching we found her three fields away, lying at the feet of her clever old mother being licked all over and comforted far more ably than we could ever have done. We had not seen the two Dollys talking to each other for ages and just how young Dolly knew where on the farm her mother would be we had no idea. We were glad to see that our policy of leaving gates open to allow all the stock to choose where to roam had been vindicated. At least Dolly's slow, staggering quest had not been thwarted by five-barred wooden barriers. After six days of constant togetherness the Dollys parted again, happily, and went their own ways.

The instances where cows reject or ignore their calves are pretty rare, and in our experience they are always resolved within a short space of time. As far as I can remember, the case of Olivia rejecting her mother's friendship was unique. Here, almost every day, we see daughters consulting their mothers about impending confinements, or maybe just discussing the weather.

Recently, we were not exactly sure when young Nell was going to calve so we decided that she and her mother should spend every night in the barn during the preceding x days (x turned out to equal 9). At 4 a.m. she started to calve and her mother watched attentively. After the calf was safely delivered (this having required help from two men), Nell senior, or Gold Nell to give her her full name, came very close, head on one side, and looked at her daughter and granddaughter with great care. She decided that both were fine, and she marched towards the gate and asked to be allowed out. She had not shown even the slightest inclination to go out on any of the nights she had kept her daughter company but tonight, knowing that she was no longer needed, was different. Thereafter she maintained a very active friendship with her newly expanded family.

Jake

All of our herd bulls have been admirable and interesting individuals: Jonathan, Ivor, Tor Down Hyadal, Olé, Mr Mini, Sam and John (the identical twin sons of Constance), Wheatrig Patriot IX, Augustus and the Bishops of Gloucester and Worcester. Jake, though, was king.

Jake's whole ancestry deserves recording. Emily, Jake's grandmother, was a Hereford, red with a white face. She became ill when she was only a few months old. We are not sure why. It seemed as if some sort of pneumonia, with accompanying loss of appetite and breathing difficulties, had taken hold. She rapidly lost condition and looked very vulnerable. My father took her under his wing, and nursed her devotedly and knowledgeably, lying down beside her to warm her. He wrapped her in hay and gently coaxed her back to health. It took months before she felt really strong but gradually, trustingly, she began to thrive. Eventually, from being little and thin she turned into a stocky, robust, square, well-coated yet fluffy individual.

Emily's first calf – Jake's mother – was Nuffield. When she was born Nuffield was dark, dark brown with a white face, and for a short time we called her Emily II. After a few months, to our surprise, she appeared to be moulting and the brown hair came off and slowly but surely Emily II turned black. Her odd renaming was due to the fact that when the Leyland tractor company bought the Nuffield tractor company they painted the orange Nuffield tractors blue. After a while some of the blue paint flaked off to reveal the original orange colour; hence Emily's new name.

We nearly despaired of Nuffield ever having a calf of her own; she just would not conceive. We had almost given up hope when we decided to give her one last chance and use artificial insemination concurrently with the herd bull.

Nuffield conceived, but it was not until nine months later that we realised she had conceived to both bulls simultaneously. She produced utterly non-identical twins, Red Ruth and Black Jake. Ruth was red with a white face and Jake was absolutely black. If we had not seen them being born we would not have believed it possible for twins to be so different. Although Emily was endearingly friendly, Nuffield was imperiously independent, and the twins were a delightful mix of these two characteristics: easy-going and trusting yet very well able to stand on their own feet. Nuffield was very proud of them, and so were we.

During the months that followed we became aware of some of the very special and unusual qualities that both Jake and Ruth possessed. They were both highly intelligent, able to work out what to do in all circumstances, and capable of asking us for help when it was needed. Jake would come up to me in the field and tug my coat with his teeth to get attention.

Ruth's first calf was as tiny and thin and fragile as Ruth was by then four-square, robust and strong. Little Ruth needed endless, patient care for several months and we had to persuade her mother to go out without her into the fields to graze, because of course she was reluctant to leave her calf behind. She soon realised, however, that we would take care of Little Ruth and she developed her own routine of grazing for two or three hours then marching back to the barn to allow us to take some milk from her to give to the calf in a bottle. Then off she would go again. (Her behaviour presaged that of Fat Hat II, of whom more anon.) Gradually, Little Ruth gained some strength and began to accompany her mother. In the intervening time we had taught her to eat hay and had cut and carried various grasses to her. She had shown a preference for mouse-eared chickweed which we sought and brought assiduously.

Jake soon became the most important animal on the farm – not in his own opinion, however, for unlike most bulls he was not at all conceited. He was magnificent: totally black, rough coated in winter, smooth and silky in summer, always with tightly curled hair on his forehead. He had neat, strong, black feet and kind, intelligent, knowing eyes. We all loved and admired him, as did the entire herd. He was gentle and never bossy, although three times stronger than everyone else. Even a smallish animal could push him away from a flake of hay. (There are normally seventeen flakes of hay in a bale, weighing about eight pounds each.)

Jake trusted us. We never disappointed or worried him and he was a happy soul, regularly talking to his mother and, by now, three sisters. (Nuffield had produced twins again: Augusta and Octavia, who were identical heifers, black with white faces just like Nuffield herself.) We decided to keep Jake as our herd bull. He was ideal, easy to handle and perfectly safe.

One day I needed to move him right across the farm from the group he had been with all winter to another, larger group half a mile away. As it would have been a pity to uproot the whole group just to move one animal, I gently pushed Jake away from his herd in the direction of the gate that led into the wood. He turned to look at me, questioningly. I patted him more firmly and he moved forward. Through the dark wood we went, with him trusting that I would never take him anywhere he would not actually prefer, and me tapping and pushing and speaking words of praise and persuasion. We reached the gate into the paddock and Jake, polite as always, waited for me to open it, then walked heavy-footed through the stream and up the very muddy bank. He turned to question me again and I reassured him that he would be glad when we got there, accompanying my words with firm taps and pushes.

Our long and interesting walk across the farm was pleasant but unsure, because although Jake lumbered slowly, even painfully at times, giving an impression of great age and boredom, I knew he could turn on a sixpence and disappear into the distance in a split second if he felt like it. Eventually the other half of the herd came into view and Jake turned to me once more to acknowledge that he understood the purpose of the walk, before hurrying to join his new friends.

Jake did have one vice. It was not, however, a vice normally associated with bovines: he loved sniffing the carbon monoxide fumes from the Land Rover exhaust pipe. At first, we did not notice what he was doing. We were accustomed to driving into the fields, loaded to the eaves with bales of hay; ten at least inside and two or three tied on the roof rack. If it was a cold day, as so often, I would leave the engine ticking over while I leapt out to spread a bale, trying to dodge the eager heads and horns and feet. Then I would jump back in and drive forward, repeating the trick at intervals and making sure the hay was fairly distributed over a largish area so that the smaller and more timid animals had plenty of flakes to choose from and were not intimidated by those more self-assured. Jake would see us coming, stroll over to the back left-hand corner of the Land Rover and breathe in the fumes in ecstasy. We realised what he was doing only when one day in his enthusiasm he started to rub his head on the bumper while still breathing in the fumes. He seemed to get carried away and the Land Rover began to rock from side to side. Our verbal remonstrations were to no avail and when I got out to persuade him, physically, to stop I saw what he was doing. After that we always turned the engine off, however cold the weather.

Unusual behaviour needs investigating

The original Fat Hat made a lasting impact on our lives. She was a Beef Shorthorn, and was originally named Harriet. When she was young she grew quite fat and was nicknamed Fat Hat. We were keen to increase the number of Shorthorns we had and hoped that Fat Hat would give us some daughters. She gave us nine sons in a row. Her tenth calf was a beautiful strawberry roan daughter we called Bonnet and she herself gave us eleven calves. Fat Hat's next calf was a blue roan bull, The Blue Devil. He was a very selfish, bossy calf and the only one she did not love. She looked after him of course but was visibly relieved when he went off to play with his friends. When he was tiny he was just another calf but as soon as his character started to develop it became obvious that he was not only an independent individual but that he was difficult and demanding too. Her first nine boys, all red and all friendly to her but not to us, had all been called Ronnie. Fat Hat's twelfth calf was another strawberry roan heifer called Straw Béret and her last calf was yet another roan called Fat Hat II.

Fat Hat was a remarkable cow in many ways. She preferred men to women and she did not like men much. She never needed or asked for anything: no help to calve, no extra food. She was never ill. In fact she was nearly twenty before she needed us.

I found her one warm summer's day in the yard when all the rest of the herd were a long way off, grazing. Unusual behaviour needs investigating and as I approached her I could see that she had plain fencing wire wrapped round all four feet. My immediate thought was that even if I had a team of helpers she would be very difficult to restrain and would probably struggle and tighten the wire. I was on my own with no immediate prospect of help, yet she seemed to be asking me to come to her aid – the first time she had ever needed a human being. I bent down, talking to her as reassuringly as I could, and took hold of a piece of the wire. She did not move a muscle. I began to untwist it, aware all the time that she might kick out. It was as if she knew that I could succeed in this difficult task only if she were totally cooperative. She stood perfectly still while I unwrapped and unwound, straightened and removed piece after piece of wire. I had to lift each of her feet in turn and out of turn. It took quite a long time but eventually she was free. Before marching back out into the field she turned and looked at me. I like to think it was a sort of grudging admission that humans do occasionally have their uses.

A little bit on names and more on grieving

All our animals have names, of course, and many have nicknames. Often the nicknames become so dominant that we forget the original names. Terms of endearment aside, the first of this group I remember was Highnoon IX. She had been a solid, unremarkable cow until she produced an unusually marked red-and-white bull calf who bumbled about instead of walking, in a cross between a stroll and a lumber. Bumble became quite an eye-catcher and his mother became known as Mrs Bumble and the Bumble suffix endured. Just over a year later Mrs Bumble had a heifer calf, Miss Bumble, who in turn had twin heifer calves, the Misses Bumble. Eventually and by an extraordinary route the first Miss Bumble became known as Granny.

When the Misses Bumble were nearly three years old, they both produced red bull calves. To recognise their new status we renamed them Mrs Ogmore and Mrs Pritchard after the characters in Dylan Thomas's *Under Milk Wood.* It was the long, not-too-cold winter of 1989 and we had a whole bunch of young calves: Amelia, Dreamy, Eleanor, Ninette, Horatio (Nell's son), Laura, Edward . . . Every day, depending on the weather, these calves went either up to the L-shaped field (their favourite) to graze or down into the Poplar Wood. They played by the stream and among the young ash trees and fallen poplar branches and by one uprooted tree, which provided endless fascination.

Although the cows and calves spent the nights in the barn where there was ad-lib hay to eat, we always took extra hay out into the fields during the day because, although they grazed the grass, in winter it has very little goodness in it. One day as we drove across the L-shaped field to distribute hay, we saw, to our horror, Mrs Pritchard lying dead in the middle of the field. Pritchard, her three-week-old son, was standing by her in bewildered amazement. The post-mortem revealed that death had been caused by an abscess on her liver, probably the result of a knock she had received, unnoticed, months or even years before, which had suddenly and unpredictably burst.

For many years we have observed the depth of emotional and physical attachment that cows and calves feel for each other. We have noted that a cow grieves for a dead calf far longer than a calf for its mother. It is possible to diminish the grieving of both. It is important for a cow to be able to 'talk to' other members of her immediate family. It also helps if we provide different, tempting, food. A calf also needs to communicate with relatives and friends. Extra attention from us in the form of food, grooming and sometimes the distraction of a change of surroundings can help to speed the process of forgetting. A lot depends on the age of the calf and we have developed different strategies for different ages. When Lotte's mother died, her older sister Charlotte took Lotte under her wing so devotedly that the calf accepted the new situation surprisingly quickly. Pritchard, however, did not appear to have made any particular friends, although he was definitely one of the gang. A young calf's main preoccupation is with hunger, but the older the calf the more it misses its mother. With much cosseting and grooming and handling and spoiling, however, even a six-month-old calf will appear to have forgotten its mother within a week, and often within three days.

Pritchard was much more hungry than lonely but it took a great deal of canny and patient persuasion to make a bottle of milk acceptable to him. By the third of his small, frequent feeds, however, we were already firm friends and I could see the next ten months mapped out before me in terms of warming bottles full of milk, long conversations and plenty of brushing and combing.

We decided to leave Pritchard with the herd. Although he would be dependent on us we were determined for him to keep his identity as a member of the group. Often it would be almost midnight when I trundled up the hill with his last feed, champagne bottles of warm milk clanking at my sides. Pritchard never took a step towards me, but stood stock-still wherever he was and waited for me to find him. Sometimes it was easy. Sometimes every other red bull calf in the field seemed to be trying to impersonate him: Ogmore, Jack, Horatio all stood quietly, allowed me to offer them milk in the midnight dark and, with what I fancied were amused smiles, steadfastly declined. Sometimes Pritchard would be standing behind a tree and there he would stay until I found him. I am sure he was not hiding, but merely waiting, absolutely confident that I would be able to locate him.

When Pritchard was four months old, the original Miss Bumble (Pritchard's grandmother) calved and had a tiny, black daughter we called Dot. Miss Bumble had more milk than Dot could drink and so she decided, without any encouragement from us, to adopt Pritchard. Pritchard's 'Granny', so we believe, knew who he was, having presumably been introduced by her daughter before she died. She was dry at that time and could do nothing practical to help. As soon as she had calved and was producing milk she invited Pritchard to be brought up with Dot. Dot and Pritchard became inseparable friends and Granny was a wonderful mother to them both.

My job as stand-in mother continued part-time for a while and I offered milk twice a day for the next fortnight but Pritchard made it clear that he could not drink another drop. He had, though, become addicted to being groomed, and although Granny licked him daily he still liked me to brush him.

A brief note about sleep

All the words and phrases we humans use for sleep – dozing, catnapping, resting, having a bit of shut-eye, and so on – could equally apply to cows and sheep and hens and pigs.

If animals feel totally relaxed and safe and know themselves to be in a familiar environment, surrounded by family and friends, they will often sleep lying flat out. They flop in a variety of often amusing positions and look anything from idyllically comfortable to dead. The sleep may sometimes last only a very short time, but we feel that it is important and that they should not be disturbed. It might sound eccentric to suggest that the reason an animal is bad-tempered is because it is short of sleep but as sleeping is vital, deprivation will obviously do harm. The animals can make up for deficiencies in their diet by foraging and finding what they need. It is up to us to provide conditions in which they can be comfortable and happy enough to sleep well.

Some of our cows have horns. Some grow downwards; some grow straight out to the sides, often achieving great length, and some grow uphill like pitchforks. I am not sure how they do it, but when cows decide to sleep flat out, none of them let their horns get in their way.

Often a cow or calf will appear in every limb, breath and eye to be sleeping soundly but a telltale, radar-dish ear will appear and rotate slightly, monitoring and analysing every footstep, creak and groan. If reassured, an even deeper sleep might follow – chewing jaws stop moving, whiskers stop twitching, and watchful, sleepy eyes disappear from view. Two minutes pass or three and the antenna scans the airwaves again. Up and down and round, this instinctive agent of survival seems programmed to receive sufficient advance warning of impending harm.

If cows are in a group of animals they do not know very well, they may choose to catnap in a seated, folded position, ready to wake and perhaps jump up at a second's notice. Or there may be occasions when they are a bit squashed and have tucked themselves up in a small space, next to a friend or relative, happy and secure but with no room to lie flat. In these cases, contentment shows on their faces and their eyes refuse to stay open before they fall asleep almost unwittingly.

There are lots of in-between examples too. Much depends on whether the animals are in a barn or out in the field. If there is a strong wind out in the open, and even if they do not feel at all cold, they will probably elect not to lie flat out but might adopt a curious chin-on-knee position or curl their heads round as a pigeon would. Many of the bovines here use their friends as headrests. Some of the older cows will gather their children and grandchildren around them at night, especially out in the open air.

Pigs, if given a comfortable, well-equipped home, will burrow to comfort or abandon themselves to sleep in an enviable way. Sheep can be as varied as any animal but I have never seen them quite flop to the extent that cattle

do – this is possibly because they sense the danger of becoming cast, a farming term for an animal lying on its back, unable to move. Unless the animal is found in time this can be fatal. At night, hens often tuck their heads into their wing feathers in true birdlike fashion. During the daytime they can appear to be asleep when enjoying the extraordinary ecstasy of sunbathing. They spread their feathers wide to expose the maximum area to the sun, tip drunkenly to one side in order to stretch one leg and, in this uncomfortable-looking, stiff state, proceed to enjoy quite prolonged sessions of rigid relaxation.

If any sense of fear is aroused then none of the animal kingdom will sleep. If the source of fear is remote, cows will communicate to very young calves that they can sleep while the older animals keep guard. At least, this is how it seems to us after decades of observation.

The security and stability provided by extended family groups has far-reaching benefits. In stressful conditions where there is competition for food or lack of space or too few water troughs, it appears far more likely that an animal will succumb to illness. The converse obviously means that living at ease with access to food and water and the reassuring fact of having friends and relatives always on hand seems a fine bolster against ill-health.

Different kinds of mooing

One thing that has always seemed of interest to the various students and farm workers here over the years is learning to distinguish between different kinds of mooing. After relatively few lessons or descriptions some have taken pride in relating incidents where recognising the importance of a certain type of mooing had resulted in good decisions being made. Cows moo for various reasons and sometimes for no (apparent) reason.

One day a man who had been with us only for a short time came rushing to the kitchen to tell me that a cow was mooing in a very loud and agitated way and asked me to go with him to ascertain what was wrong. To have left the kitchen at that moment would have resulted in burned biscuits so I forced him to delve into his hitherto untapped powers of description to describe the cow in sufficient detail for me to know who it was and to describe the mooing more precisely. I judged that she was not distressed but merely cross at having temporarily lost sight of her calf. I gave him a foolproof visual description of the calf and he was then able to reunite the pair.

As I have already said, cows moo for various reasons: fear, disbelief, anger, hunger or distress. Each cow, moreover, has her own method of asking a question, either with a look or a strange, quiet moo.

Sometime after midnight on a very cold night in February I was wrenched from a deep, tired sleep by a cow mooing. It was not a moo of annoyance or boredom; it did not signify hunger or pain. It was a moo of absolute determination. Determination not just to wake me up but to make me get up and go out, and immediately. I did not know then that it was Araminta. I simply knew that I must hurry both for her sake and to save the rest of the family from being woken.

I grabbed my towelling robe and plunged down the back stairs by feel, reached the basement, pulled on my Wellingtons, no time for socks, realised it was raining, snatched my mac and ran up the steps and along the path, trying to button up and feel my way between the bushes simultaneously. Araminta was still mooing. I wanted her to stop mooing more than I wanted a torch so I headed for the noise. It was very, very black dark. Saying kind, solicitous words, I discovered by touch who she was and that her udder was full of milk. I guided her towards the cow pen, hoping that she would stop calling as soon as the pressure in her udder was alleviated. I knew as we walked that her son, The Don (in memory of Sir Donald Bradman) must either be ill or dead, because her milk had not been drunk all day. I knew, too, that her penetrating moo was not just to tell me she was uncomfortable but that I must do something to help her son, since she could not.

When I had fed and milked her, I picked up a torch and explained that she must show me where The Don was. When our cows do give us credit for intelligence they tend to make the mistake of presuming that we know everything, and after I had opened the cow-pen door and pushed Araminta out, she just stood still. I was afraid she might start mooing again to remind me to find her son, so I pushed her in one direction, in a fairly businesslike way, as if I knew where we were going. Trustingly she complied with my instructions. After fifty yards I stopped. She stopped. I gently turned her round and pushed her back the way we had come. This made her realise at once that I had not got a clue where we were going and she turned herself round and marched off in the original direction at twice the speed. I followed.

We found The Don three fields away. He was standing up but looked very sorry for himself. He was severely 'blown', and in such circumstances death can be sudden. He was reluctant to move but I forced him to walk home, his mother walking by his side. When we reached the yard, I put him in the cattle crush to keep him still and fetched the long rubber tube we kept for such eventualities. Opening his jaw with my left hand I gradually eased the pipe down his oesophagus and into his stomach. Holding the end of the tube in my right hand I massaged the left side of his abdomen with the other until eventually the trapped air was released.

Although I cured The Don for that night and he and his clever old mother snuggled down in the straw in the barn, his problem recurred. With veterinary approval the same course of action was repeated time after time but it finally became necessary for the vet to perform a small operation. In the end The Don did make a complete recovery but only after almost two months. Strangely, he appeared happy throughout his ordeal: when in discomfort, he stopped eating and drinking, but did not seem distressed and only minutes after being treated he was eating and drinking again as if nothing had happened.

Cows make good decisions

As I have already said, Old Fat Hat's first daughter was called Bonnet and Bonnet produced eleven calves, of whom Roan Bonnet, Little Bonnet, Peter Bonnetti, and Gold Bonnet are notable. Roan Bonnet's first calf was called the Bishop of Durham. For some reason she did not take motherhood all that seriously and although she did most of the right things – feeding, grooming and staying with him – she was a bit half-hearted, decidedly not over-protective and produced only a small amount of milk.

Consequently, Durham was psychologically balanced but rather small and slow-growing. We decided that he needed some extra food, and since it was late autumn and there was no grass to graze worth speaking about, once a day Durham was given a feed of home-grown barley and a couple of hours away from all the other cattle to eat hay at his leisure. He very soon learned to tell the difference not only between men and women but also between two similarly sized men. He never asked the same person twice for food on the same day but if a different person approached him he would try pretending for all he was worth that he had not been fed that day. Quite often this procedure worked. After Durham, Roan Bonnet had the Earl of Warwick, and then the Duke of Lancaster, whereby hangs another tale (see p. 74).

We have found over the years that if they are allowed the right conditions to live in, cattle make very good decisions. They need access at all times to shelter, pure water and good food, freedom from stress and a level of stability. If the weather forecast predicts rain but the cattle insist on staying out on the exposed pastures in midwinter, or if in the middle of June with weather supposedly set fair, they come down to the barns and ask to be let in, we do well to take heed. One June evening they all came crowding in just before a torrential storm that took everyone else by surprise. A few hours later, at 1.30 in the morning, it became dry and very mild and they were determined to go back out to graze. They made their wishes known vociferously enough to wake all the villages around. We dutifully and hurriedly got up and were taking them down the road to some more interesting pasture when we were stopped and questioned by an amazed policeman. We were tempted to say that we were merely taking the herd for their usual late-night stroll.

I have talked about the importance of ensuring that animals have permanent access to shelter. All sorts of structures can play a part: trees, banks, walls and barns each have a role. However, the most important and versatile living shelter is a hedge.

Many writers past and present have written lyrically about the multi-purpose hedge. Every old hedge has a story to tell, it being possible to calculate from inspection when a hedge was planted. It is also possible to deduce why it was planted where it is.

The madness of removing hedgerows cannot be overstated. The many thousands of miles that have been destroyed have resulted in far more than mere visual deprivation: gone too are roses of amber, white and several shades of pink cascading from the topmost boughs in May and June, resulting later in berries of every hue. Innumerable wild species rely on these 'little lines of sportive wood run wild', as Wordsworth called them. Birds have skyscraper dormitories and layered nesting sites, not to mention their life-sustaining winter larder. A hedge that is old enough provides rosehip, plum, elderberry, crab apple, haw, nut, sloe, acorn, ash keys and honeysuckle berries. Timid creatures find safe havens, nipping easily through the barbed blackberry skirts to escape the fearless badger. Rabbits, dormice and field voles all find their hidey-holes and the vulnerable ground-nesting birds have an enhanced chance of survival.

On this farm, two parallel hedges were planted on either side of a farm roadway ten feet wide. By the time we came to live here the hedges were thick and strong and about twelve feet tall. Today, over thirty years later, they are thirty to forty feet tall and arch over the road, meeting in the middle to provide the appearance and convenience of a barn. All the cows know where this 'barn' is and will choose to use it both in winter for shelter from cold and in summer to escape the heat.

Coping with rain is part of the syllabus but it is not always easy to get it right. A group of mature cattle can stand an enormous amount of rain and, in fact, seem not to notice it. In a mixed group with very young calves, some of the less experienced mothers do not always realise that their offspring need more protection than they do, but the clever, older cows always know and bring their calves into the barns or find a sheltered spot under a tree or close to a hedge.

Some fully grown sheep, with their weatherproof wool, actually like cold wind and sometimes walk to the highest, most exposed, part of the farm to revel in it, but they do not like prolonged rain. When they have lambs at foot, more care is needed. Lambs can be quite vulnerable during the first few weeks of life and although most can cope with cold or wet weather, the process of coping requires much of the energy they are deriving from their mothers' milk. Providing some sort of shelter is, therefore, financially as well as humanely desirable.

As a general rule we have found that lambs are even cleverer than calves at finding comfortable places to shelter. Very young lambs will commandeer a flake of hay or nip inside a shed or any makeshift shelter. If nothing suitable is to be found they will try to improve their lot by any means they can think of. This might involve climbing on a tree stump or log of wood or into a hollow tree or onto their mother's back in order to be dryer and more comfortable than they would be on the ground.

The decision-making process animals are constantly involved in includes choosing exactly what to eat. Nibbling and browsing all sorts of different grasses, herbs, flowers, hedges and tree leaves gives them vital trace elements in their daily diet in the amounts they feel are appropriate: such decisions could not be made so effectively by us.

The animals are all individuals. Mass 'legislation' for the entire herd in terms of feed might suit the majority but we have always been concerned with minorities. It is not only more accurate and effective to let the animals decide; it is also cheaper. We have watched cows and sheep eat extraordinary plants in prodigious quantities. Cows will eat dark green, vicious-looking stinging nettles by the cubic yard and sheep often choose pointed, spiky thistle tops or tall, tough dock leaves, particularly after parturition when their energy reserves are depleted. And this is when they have access to good natural grass.

I have never known a sheep to take a day off but cows do fairly often give eating a miss for a day or so after giving birth. In my experience new sheep-mothers always eat faster and more single-mindedly than ever before, knowing that they must produce enough milk to satisfy the relentless demands of their offspring. Some cows, if they have produced a smallish calf, will know that they have milk 'in reserve' and might spend the first day or two under a tree with the calf, taking it easy. They might graze a little in the evenings but nothing like their usual daily routine.

One particularly satisfying fact we have discovered is that if the animals have sustained an injury they like to eat quite large quantities of willow. We hope that this is connected to the origins of aspirin. If a willow tree is not growing in a handy place we cut and carry boughs to whoever needs it. Without exception they eat keenly, sometimes on several consecutive days. When they feel they no longer need it they will just walk away.

Desdemona II, granddaughter of Dizzy, or Black–and-White Desdemona as she was generally known, preferred grass to every other type of food. All was fine in summer when she was born and the autumn and again in spring but in the winter, no matter how bitter or even icy the weather, when all the other cattle were pleased to stay in the barns, she would stand at the gate and stare until we realised what she wanted and opened it. Given carte blanche she would plod off maybe half a mile and graze quite alone all day, though just occasionally she would pass within speaking distance of the sheep who also prefer to stay out in most weather conditions. She simply would not eat hay or straw or barley or apples; she wanted only grass. She could not and did not get fat on grass in winter but she was very contented and late every afternoon she would come slowly home and ask to be let back in to spend the night with her mother, sister, grandmother and cousins. By her second winter she had learned to eat hay but she still preferred grass and spent most of each day grazing, whatever the weather.

Very young calves will 'stay in bed' if they are not feeling well. Even if their mothers go out to graze and call them to follow, many a calf will just stay where it is, and the mothers always come back at regular intervals. One of the canniest examples of calf good sense was the case of Chippy Minton.

Chippy's mother was inexperienced and had taken him out in the fields on a December morning. But Chippy decided that he would be better off in the warm barn.

At six days old, he left his mother and the rest of the herd and plodded home, a distance of nearly half a mile. Luckily we were there to see him descending the hill and were able to usher him into suitable accommodation. He had caught a bit of a chill and was scouring, so he needed rehydration therapy and nursing. It was at this time that he developed a penchant for being groomed: he hated to go to sleep at night with muddy legs.

Bovine friendships are seldom casual

It is extremely common – the norm in fact – for calves to establish lifelong friendships when only a few days old. Sometimes three calves all born within a short space of time form a group but more often it is a two-calf friendship, usually between the two who are closest in age. The White Boys – pictured on the front cover – are a case in point.

Nell calved first and produced a pure white bull calf, a dazzling white. Juliet calved the very next day and produced an identical calf. We had never had such white calves: grey, cream, buff, off-white, silvery, golden but not pure white. The first calf walked over to greet the new arrival and stared at him as if looking in a mirror. They became devoted and inseparable friends from that minute. The two mothers, now of secondary importance in the lives of their two offspring, became firm friends too, forced as they were to spend all their time together, waiting around to provide milk on request. Nell and Juliet had both had their own childhood friends but their new circumstances threw them together. A year later, when both mothers calved again, there was absolutely no jealousy, no trauma. The White Boys had each other and barely noticed their mothers leaving the field to come down nearer the house to the nursery field. Nell and Juliet remained friends, this time with relatively run-of-the-mill calves, one red and one whitish.

The White Boys lived in a world of their own; in the midst of a large herd but oblivious to it. They walked round shoulder to shoulder, often bumping against each other, and they slept each night with their heads resting on each other. They were magnificent: tall, gentle, independent, kindly, though not over-friendly, noble. One had a pink nose, one a grey.

A less eye-catching but equally strong friendship had been forged six years before between Black and White Charlotte (daughter of Charolais Charlotte) and Guy (son of Dizzy). She was all black with a white tail and he was a smart dull grey with a white tail. Guy was a member of the Discount family and Charlotte was a Highnoon.

Bovine friendships are seldom casual. Devotion is the order of the day, although it is directed towards practical mutual help and is rarely over-protective or emotional.

Charlotte and Guy got on like a house on fire but they each had other friends. If they were sometimes parted by the grazing preferences of their mothers, there was no pining but there was always a joyous reunion when the split herds grazed their way back together again. (Anne and Helen, I remember, actually kissed each other after an unplanned one-week parting when they were both aged three months.) The friendship tailed off when Charlotte's mother died and she took on the role of mother to her little sister Lotte. Lotte was a smaller version of her sister, black with a white tail, but she had horns whereas Charlotte was naturally polled. Charlotte was so caring and kindly that Lotte coped with this trauma almost unscathed, and when Charlotte had her first calf Lotte was an invaluable aunt.

Charlotte, however, did not take naturally to real motherhood. Her adorable all-black daughter, born on the Ides of March and named Calpurnia after Caesar's wife (and almost immediately nicknamed Cocoa), was not permitted to suckle milk from her debutante-type mother who announced straight away that the nanny could bring up the brat. 'Nanny' was my brother and he patiently haltered Charlotte and persuaded her to stand still while Cocoa drank, taking her kicks of protest on his own shins to protect the calf.

Cocoa loved her mother (she never actually got kicked, which might have taken the edge off the relationship).

She suckled milk three times a day, was loved and stroked and groomed by all of us, was treated very nicely indeed by Lotte, and thought that life was wonderful. Mothers, she must have thought, give you milk if forced to and people give you affection and protection.

Charlotte did not spurn her daughter's loving overtures; she just seemed not to notice. Cocoa would nuzzle under her mother's chin, occasionally giving her a playful bunt. Charlotte just gazed into the distance. They always grazed and slept very close together but that had nothing to do with Charlotte.

Cocoa was a real beauty and everybody's favourite. She loved everyone and welcomed attention. Even the men who worked on the farm, not celebrated in those days for their overtly affectionate natures, never failed to pause for a quick stroke, however busy they were.

When Cocoa was just over two months old, everything changed. She risked sneaking a drink when we were not there; Charlotte noticed her for the first time and obviously thought how splendid she was and she told us in no uncertain terms that she intended to take full credit for her lovely daughter, and from that day on she licked, polished and fed Cocoa single-handedly.

This friendship was of particular interest to us because Cocoa was so used to people and so friendly that we could be part of the process without imposing.

When Charlotte had a bull calf, Cassio, fifteen months

later, Cocoa was on hand to help. She stood over him and protected him from other curious cows and she babysat happily whenever the need arose. Even when Cassio was grown up and calf number three, Carline, was the centre of attention, Cocoa and Cassio would often be seen standing sharing the same flake of hay. Charlotte, needless to say, never forgot again how to be a good mother; in fact she was one of the best.

But bulls are a completely different kettle of fish

Although living with cows is a rewarding and always interesting experience, bulls are a different kettle of fish altogether.

At one time we had three bulls of roughly the same age: the Bishop of Gloucester, a Welsh Black; the Bishop of Worcester, a Lincoln Red; and Augustus, a Charolais. At one stage it became necessary to remove Gloucester and put him with a particular group of cows near the house. Meanwhile, Worcester and Augustus rubbed along well enough. I think they liked each other and they certainly never fought, but Augustus thought himself superior and Worcester was happy to let him get away with some prima-donna displays. After the bulls had been parted for several months an incident occurred.

We had a young student helping on the farm. He had been given a lot of jobs to do and had had it impressed on him in unequivocal language not to open the gate to the enclosure where Gloucester was: he must not be allowed to wander. Sometime later, the student came hotfoot to the house revealing between gasps that Gloucester had taken the opportunity to escape during the few minutes he had judged it safe to leave the gate open. Such a thing had happened before, when an actor friend had offered to help and had gone in pursuit of the escapees only to land on his face in deep mud (hence our stringent orders). I raced up the path, in my slippers, shouting orders behind me for him to find my brother and follow.

I knew that this time it was more serious, and I knew where Gloucester had gone: Augustus had been growling in a deep, menacing, half-bored, half-grumpy voice all morning and Gloucester had been listening. Gloucester had gone down the steep track and into the paddock. I jumped into the Land Rover and sped down the farm road, plummeting down through the trees faster than ever before. The bumping, shaking and jolting threw me about all over the place, and I dared not even contemplate the possibility that I might get stuck on the waterlogged bridge. Hurtling through the gateway into the wood I managed to park in front of the gate to the bulls' field literally a second before Gloucester galloped up.

Augustus pawed the ground on his side of the gate, Gloucester champed at the bit on his. Worcester was shadowing Augustus's every move but without the least conviction.

If they are evenly matched, bulls will fight all day and all night until they are utterly exhausted. There is never any trouble if they grow up together as peers or if a young bull is introduced to an older one when very small and they then stay together. But if bulls are parted and then reintroduced, the earth trembles and, short of using tranquillising darts, humans are useless.

I knew I had to keep the two of them apart until reinforcements arrived. I paced up and down, growled when they growled, threatened when they advanced, consoled when they retreated. I had never felt more determined. The consequences of failure would be too hard to bear.

The two men arrived an eternity, or five minutes, later and we set about walking Gloucester home. We threw some hay from the back of the Land Rover to divert Augustus's attention and we all three guided Gloucester with wills of steel, sticks of ash and howls of unimaginable inventiveness – driving cattle is an art in itself worthy of deep discussion and at the best of times it is more reliant on psychology than strength.

Gloucester trotted forward, overwhelmed and taken off guard by our forcefulness but after fifty yards or so he made a skilful attempt to double back. He was very excited, very strong, very fit and very fast. Somehow, our do-or-die attitude deflected his bid and we progressed a bit further. We kept talking to him, admonishing him, threatening him, telling him he should obey us if he wanted to stay alive. Gloucester did not believe a word, but gradually, with him stopping and starting, galloping and braking, snorting then trying to look pitiful, we inched him towards home. We only just made it. Our every sinew, both mental and physical, was strained to the utmost. He could easily have outrun and outwitted us, but perhaps his inbuilt inclination to trust us, coupled with the strangeness of our frantic gestures, enabled us to get him into the barn. We relaxed, laughed, collapsed, ecstatic, worried still but triumphant. Short-legged as he was, little Gloucester suddenly and effortlessly cleared a five-barred gate and was off to the wood again. Another gate was closed in front of him and this caused him to hesitate just long enough to allow the recapture.

The bulls never met again. Gloucester had his herd and the other two had theirs. Vigilance reigned and harmony returned.

As I write it is still only March and although spring has popped up in places, with primroses everywhere, wood anemones somewhere, violets few-where and oxlips only one-where, it still gets chilly at night and I must put down my pen and walk to the top corner of the Cherry Tree Field, surrounded by invisible but hooting tawny owls, to see if the youngest calf on the farm has found a comfortable and sheltered spot for the night.

Fat Hat II

All animals are individuals. Some will impress their characters on you, while some will glide through life keeping a low profile. The better you know an animal, the more use you can be to it. If you know how it is likely to react in various circumstances you can be prepared. If you observe how it communicates everyday needs you can interpret unusual situations far more effectively. You can also learn which herd members can be left to their own devices.

The observations we have made quite often have no relevance to the everyday lives of the animals. It is likely that even more interesting and significant things occur when we are not there at all. However, such observations as we have made have given us a heightened awareness of our animals' intelligence and an ability, sometimes, to anticipate events and therefore save suffering and, on occasions, lives.

Fat Hat II's usual philosophical, kindly, gentle, intelligent and trusting nature was sorely tested and in fact temporarily changed by the events surrounding and succeeding the birth of her second calf.

Her first calf, the Duke of York, was an uncomplicated chunky boy with few distinguishing features apart from the short legs he had inherited from his mother and the fact that, unlike most bovines, he lapped water like a cat. This odd characteristic had been observed by us only once before when Print, a pedigree Ayrshire, always took ten times longer than any other cow to drink her daily requirement of water, lapping in leisurely fashion.

Print, normal in all other respects, and mother of two bull calves called Victor and Feather, took a strong dislike to the little woollen hat worn persistently by one of the men. She would approach him affectionately, allow herself to be stroked, then, when she saw her opportunity, she would neatly remove the hat with her mouth, dropping it carefully on the straw. However many times it was replaced she would match them with her patient removals. She never tired of the game. He resolutely refused to change his hat; she never removed anyone else's.

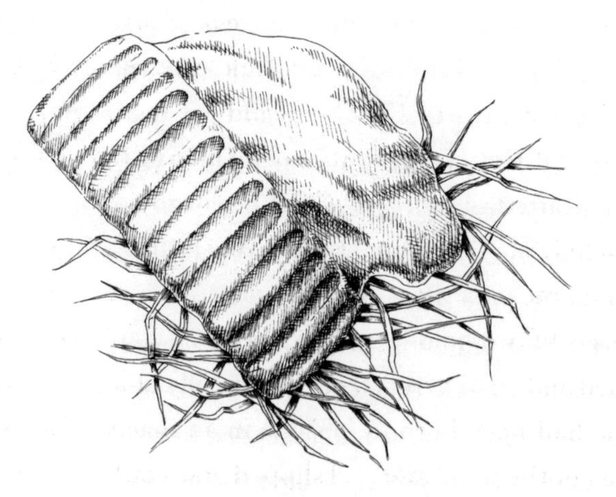

When Fat Hat II was due to calve for the second time, for some unaccountable reason she went off unnoticed into the wood, and by the time we next visited the field where she had been, several hours had passed and she was back grazing as usual. We nearly failed to realise she had calved – she looked fine, not even thin, but we still vaguely suspected something was different. We tried to persuade her to tell us if she had calved and if so where the calf was, but she pretended not to understand. We tried various tactics that had worked with other cows, but she seemed strangely resigned, distant, uncommunicative. We brought her home to examine her and when we found that she had indeed calved, we knew it was time for her to tell us more. We took her back to the field and tried to make her retrace her movements. Backwards and forwards we went, round the field, through the wood, becoming more and more concerned. We began to suspect the calf must be dead and that she had come to terms with the fact but we could not rest until we were sure. Fat Hat II simply would not help us to find the calf. We drafted in more help and began to comb the wood (60 hilly acres) systematically. When the calf was finally found the poor little mite had gone into a deep sleep. She was cold, wet and hungry. Fat Hat II, having obviously tried and tried to reach her, had finally abandoned hope. She had been born on a slope in an inaccessible, steep part of the wood and had slipped down out of reach.

Fat Hat II had hung around during our search and as soon as she saw us approach the right spot she came over so that we all arrived together. We picked up the little bundle and walked home. Fat Hat II followed closely; she was very perplexed. She was pleased to have her calf back but could not understand why she was unable to fulfil the role of mother. The calf was too frail to be able to stand or suckle from her so Fat Hat II stood by as onlooker while we warmed and dried and fed and nursed her daughter. Like an understudy in the wings she knew her role but was not called on to play it. We took milk from her and gave it to the calf in very small and frequent feeds. Fat Hat II sometimes seemed concerned and loving, sometimes feigned indifference, but she regularly showed us a sort of baffled gratitude.

If Black Hat, as we called the new arrival, had been human she would have been both frail and pale but as she was an enchanting blue-black roan being pale was not an option. She must nonetheless have felt pale, and for ten long weeks we nursed her. She was dainty, delicate, featherlight, intelligent, responsive and uncomplaining throughout. It took us three days to persuade Fat Hat II to go off and graze, having told her she had to eat in order to go on producing milk for her calf. On day two we offered her a chance to go for a walk but if she took ten steps forward she immediately came hurrying as many back to check on her daughter. We used more persuasive tactics on day three and as soon as she reached the field the wonderful taste of the grass preoccupied her completely. She stayed away for two hours then came back on duty. Gradually she developed a routine consisting of sustained hard grazing and popping back to see the patient.

Black Hat had contracted double pneumonia during her unhappy time in the dark, cold wood and her recovery seemed to take an eternity but gradually she became stronger and learned to suckle from her mother. Fat Hat II longed to take her daughter with her when she went out into the fields. She talked to her about it, called to her and did everything she could think of to persuade her to come out. But Black Hat knew she was still too weak.

One warm, windless day we popped Black Hat into the Land Rover and took her out into the fields. We drove up to Fat Hat II and lifted her daughter carefully out in front of her. Fat Hat II was so pleased. She mooed lovingly, proudly, licked her cursorily, walked round and round in dizzy disbelief, then she thanked us. And then the trouble began.

We knew Black Hat needed to be with her mother to learn to graze and to become gradually accustomed to and integrated into the herd. She was, however, still very fragile and we also knew that the process would be far from straightforward.

After less than an hour on the first day it began to rain. We flew up to the field, collected Black Hat and brought her home. Fat Hat II was furious.

Unfortunately the weather interfered often and Fat Hat II could not understand why we gave then took away her calf. Every time she saw the Land Rover, even if Black Hat was standing by her side, she would march up to the window, thrust her head right in and examine every corner. She had become suspicious of our actions and, it is painful to remember, she became very unco-operative and generally unfriendly to me.

Fat Hat II did not stop liking all human beings, just me. She still liked my mother who had, in fact, master-minded the slow-release plan and who kept watch for every spot of rain but did not physically remove Black Hat (although she was always present). Fat Hat II watched who did what and behaved accordingly.

Slowly, slowly, Black Hat grew stronger. By the time she was five months old she was a permanent member of the herd and quite independent. Fat Hat II, once she had her daughter all to herself, taught her to mistrust me. My two best friends, who had relied on me so absolutely through such difficult times, wanted no more to do with me. I was proud of them both and pleased to see them behaving normally, and I still spoke to them. But Black Hat ignored me and Fat Hat II shook her head at me angrily and if I got too close she would biff me with it. Luckily she had no horns. Even the following winter, when bad weather forced the whole herd to spend quite a lot of time in the barns, I would have to be careful walking among them. Sometimes, with a bale of hay on my back, I would suddenly find myself winded and would look round to see Fat Hat II intent on reminding me that I had done her an injury. It took three years before she forgave me.

Most stories do not have endings: I relate incidents or events and once they are over life goes on. But the story of Fat Hat II had a beginning, a life and an ending.

My mother had a special relationship with Fat Hat II; it began the day she was born. At three hours old she high-step-trotted down the field to introduce herself, leaving her mother behind (as you will perhaps remember, Fat Hat herself did not like people).

The friendship remained strong through all the vicissitudes of Black Hat I's infancy, was strong throughout the youth of Black Hat II and was still strong the day Fat Hat II died.

On the last day of a cold October Fat Hat II gave birth to a beautiful all-black heifer, this time with the assistance of the vet. She would have been called Black Hat III. Three days later, when my mother was in the barn with Fat Hat II and the new calf, she sensed that there was something seriously wrong; in fact, she knew that Fat Hat II was going to die. No one else had noticed anything wrong. The cow had licked and suckled her calf each day and was lying down beside her.

We acted on my mother's intuition and the vet diagnosed peritonitis, which he said was untreatable and incurable. In order to save her from suffering we had her put to sleep. Before she died she had, with determined and intense eyes, made my mother promise to look after her new calf. This my mother did and Fat Hat II understood. My mother never broke a promise, and this one took quite a bit of keeping, with ingenuity, determination and cunning all playing their part. Fat Hat II's niece, Roan Bonnet, had calved a few days earlier and her son, the Duke of Lancaster, did not yet require all her milk. Her milk-producing ability had improved since the Bishop of Durham was born, and her milk was likely to be similar in quality to Fat Hat II's as they were the same breed (both Beef Shorthorn) and related. We asked Roan Bonnet to adopt the orphan who was now officially known as Jane Eyre. She was not keen on the idea.

While she was tied up in the cow pen Roan Bonnet seemed happy to give us milk in exchange for barley and apples and sweet hay, so instead of a milking machine we let Jane Eyre do the milking. Roan Bonnet was no fool, however, and as soon as she caught sight of Jane out of the corner of her eye she moved from side to side to make suckling impossible. This was where Lancaster came to the rescue. We positioned him close to his mother, so that whenever she looked round she only saw her own calf while Jane tucked in behind and had a good drink. For several weeks we had to rely on Lancaster's good nature, making him stand in as a bodyguard to his adopted sister.

This was a time-consuming business, as we always had to tie Roan Bonnet up, but just before Jane Eyre was two months old Lancaster worked out a solution that benefited everyone: whenever he wanted a drink he would call to Jane to come and have one as well, even if she was some distance away. A while later he decided to take on our job of grooming her as well. Jane and Lancaster became very firm friends, and even two years later, having perhaps spent weeks apart, they would meet and settle down next to each other, looking at the view and discussing this and that.

Before this happy state reigned there was one unfortunate incident. One day out in the field, Jane asked Roan Bonnet for a drink with Lancaster nowhere around. Roan Bonnet kicked Jane. A large lump appeared on Jane's hip, growing larger daily. Strangely, it did not make her limp or seem to give her pain and she still behaved normally, playing games with Lancaster and Billy and Gulliver, all of whom had been born within days of each other. (Alfred, son of Ditch-Hog, was of course not allowed to play with hoi polloi but that is another story.)

We felt very reluctant to consult the vet: Jane Eyre's older sister, the first Black Hat, had endured so much medical attention that we did not want to embark on any medication with Jane unless absolutely necessary. Instead, we decided to try homeopathy. This subject merits a whole chapter in our lives as farmers but suffice it to say here that on this occasion it worked. Less than twenty-four hours after administering the prescribed treatment the lump had burst, seemingly without pain. Two days later the evidence of any injury was impossible to detect; we soon forgot which side it had been on, and we hope Jane did too.

Cows have preferences

A word here, while on the subject of Hats and Bonnets, about Bonnet herself. She was the first daughter, although the tenth calf, of the original Fat Hat and quite a cow in her own right. Bonnet loved apples. Most cows do, and so do sheep and pigs and birds, but Bonnet used to think about them even out of season. Whenever she saw us, her eyes would ask us if we had an apple or even a pear. During her long life she managed to communicate a number of different questions with different types of stare.

Our Laxton Fortunes and Newton Wonders stored well, on wooden slats suspended high up in an airy barn, until March and sometimes, iffily, to the end of May.

Also, space was found in a cold store, thus ensuring a constant, if limited, supply all year. So Bonnet was seldom disappointed during the months when the grass was scarce. Somehow she knew when the first of the new season's crop, the Worcester Pearmains, were beginning to ripen and she would be under the tree on the right day to reach up to the lower boughs and pick up any within reach. Five or six weeks later she would be stationed under the Lord Derby, much, much more sour apples to us, but they were all sweet to Bonnet.

All the other cows ate apples singly. Bonnet could eat four at once, effortlessly. Strangely we gave the title of the Apple Eaters not to Bonnet and her family, all of whom inherited or copied her passion for apples, but to Jacques and Maurice, an unlikely pair of friends brought together by common misfortune.

Jeanine, mother of Jacques, and Highnoon VII, mother of Maurice, both died within a fortnight of each other. We did not introduce the two boys, as both had friends of their own and both had kindly older sisters, and because there was also several months' difference in their ages which, to cattle, is usually significant. They found each other, however, and struck up a lasting friendship. We gave them both extra attention in our usual bid to help them through a trauma and one of the tactics we used was a daily ration of apples. They both learned about these very quickly. As soon as the Land Rover entered the field and drove among the cattle Maurice and Jacques would place themselves strategically to catch our eyes. Jacques would later learn to put his head through the open window so no other animal would get his allocation by mistake. Maurice was slightly more reticent and would loiter near the back of the vehicle and wait to be noticed. Jacques was a magnificent Hereford, red with a white face, and Maurice was an inconspicuous but canny Lincoln Red, but the Apple Eaters stuck together through thick and thin.

Eye contact

It is hard to write about cattle on this farm without bumping into incidents involving members of the Hat family. On 9 January 1995 Little Bonnet and her son Smasher, July Bonnet and her son JB, and Roan Bonnet and her son Red Rum were brought down to the farm buildings so that the bull calves could be counted and have their ear numbers read by the 'man from the Ministry'. This inadvertently left Christmas Bonnet, July's younger sister, with no family and *ipso facto* no friends. The next morning when my mother and I went to the Monument Field to feed the cattle, Christmas Bonnet stared very hard first at me and then at my mother, walking from one side of the Land Rover to the other and fixing us both in turn. It took quite a few minutes before we realised what she was trying to tell us and, when we did, we both apologised to her and promised to bring her home as soon as we could. This could not be straight away, though, because my brother was away all day and we had extra work to do. When Richard arrived home later that evening he saw Christmas Bonnet standing in the yard with the house cows, with whom he knew she did not belong, and staring across the road to the barns on the other side. He mentioned to us that she was there and we were able to explain the whole story to him, realising that she must have negotiated three hedges, fences or gates in order to bring herself home. Once the family was reunited they all had apples for tea.

Another bovine who achieved a lot by staring was Black Wendy II, subsequently known as Friendly Wendy. One winter she began to look a bit thinner than her contemporaries and we decided to give her some extra food each evening. She very soon learned to walk home with the milkers, but on one occasion we were delayed by unexpected visitors and her usual feeding time came and went. Wendy found a way out of the field and, spotting a man who was taking a holiday in our farm cottage, she proceeded to fix him with a stare. In fact, she followed him as far as she could and watched his every move as he pottered in the garden. He told her, as he later related to us, that although he could not fathom her wishes he would consult the appropriate authorities on her behalf. He walked the hundred yards to our house, with Wendy following. This manoeuvre resulted in the awaited food and every day thereafter Wendy, having observed everything, came to stand outside our kitchen window and stare silently until we noticed her.

Cows remember

One of the nicest attributes a cow can possess is a good memory. I say this from the human perspective of course, though I dare say a good memory is also useful to a cow. Sometimes work allocation might prevent one of us from seeing a particular group of cattle for several weeks, although someone else in the family might see that same group each day. No matter how long the parting, we are always individually remembered. Cows have their favourite people as well as vice versa. Sheep also have long and accurate memories. It seems now to be an accepted fact that they can recognise at least fifty of their individual companions. From experience I conclude that they remember all of the human beings they have ever known. The evidence I have seen indicates that they recognise us by our voices but perhaps they notice what we look like, how we walk or even our height.

A little bit about horses

In the early 1960s we had two ponies. The older, cleverer and more stubborn mare, who was plagued with arthritis, toppled upside-down into the ditch one day and was hopelessly wedged. Unusually, my father went to the field that day to catch and take the younger pony to be shod. The pony allowed himself to be haltered then absolutely refused to leave the field. Instead, he steadily pulled my father downhill towards the ditch. This extremely atypical behaviour saved the life of the old girl and we were moved and pleased to see that the younger pony had stamped the ground down almost level with the ditch in an effort to try to save his friend single-handedly.

A digression on sheep, and pigs and hens

Cattle play a major part in our lives and the more natural or wild their existence, the more fascinating it is to observe the way they conduct themselves. By the same token, our observations of truly wild animals and birds have been even more rewarding but I will continue my small digression here to talk about sheep and pigs and hens.

Audrey and Sybil, two hand-reared orphan lambs, and Gayle Elspeth Rosie, a pig known as Piggy, were quite remarkable in their own particular ways. When she arrived here in a shoebox on 20 June 1985 aged one month and one day, Piggy was just seven inches long from the tip of her nose to the tip of her out-stretched tail. The lambs, Audrey and Sybil, were both born on 22 March the same year.

Piggy had been born on an intensive farm, and bodily thrown out immediately because she was 'too small to bother with'. Gayle worked on the farm, and had rescued Piggy and reared her on baby food. She looked after Piggy in her second-floor flat with the help of her Alsatian, but she knew that Piggy could eventually weigh sixty scores and would be unable to continue her upstairs existence long before that, so she searched for a permanent home for Piggy before they became too attached to one other. I was contacted through Elspeth, a mutual friend.

Piggy did not like being taken away from her nice home and she made a one-day token protest by burrowing into a large pile of straw, from which she would not emerge (except to eat and drink when we were not looking). On the second day she decided to make the best of her new circumstances and agreed to allow us to spoil her and wait on her. She came with us in the Land Rover on trips round the farm, being gently lifted in and out and enjoying immense fun in the hay field burrowing under the neatly turned rows of hay. Even three months later, by which time she was much bigger and very heavy, she would enthusiastically accompany us on walks, even uphill, though she would insist on being carried all the way home.

When Piggy and Audrey met for the first time it was a surprising meeting of equals. Piggy was only a quarter as big as Audrey, who stared down at this extraordinary little thing with intimidating superiority. Piggy did not move a muscle. Nose to nose they stared at each other and after two motionless minutes came to an understanding that quickly grew into a strong friendship. Neither had ever known a mother.

In the morning, Audrey would frequently go and 'call' for Piggy. This often meant waking her up, which Audrey did by gently tapping her with her foot. If Piggy got up first she would find Audrey and push with her head until Audrey got up too and they would spend the day together playing, grazing and rooting under the cherry trees. Sybil would usually tag along too. The friendship lasted until Audrey gave birth to Brigitte and Lolita, with whom she knew not what to do, and Piggy produced eleven piglets, which puzzled her. Sybil produced Manuel. This was when we realised that Piggy and Audrey were both sure they were human. It took us some degree of patience and cunning to teach them to become mothers but they finally became so preoccupied with their offspring that they forgot about each other.

Audrey was a delight to know throughout the whole of her long life. Always friendly and helpful, she was also calm and beautiful. If ever we had to bring all the sheep into the barn she would walk behind with us, like a sheepdog, and if anything occurred to frighten the flock when we were not there, a stray dog for example, Audrey would bring the whole flock home via their too-narrow-for-a-cow gate. Sometimes when we went out into the field to see her, she would very politely make it clear that she was extremely busy grazing. If we approached to stroke her, she would not run away, but she might start eating with increased speed and walk away from us just slightly faster than we were walking towards her. If we actually had a reason for catching up with her, though, we could always prevail on her to stand still.

I remember once when we had invited a party of fifty eight- and nine-year-old schoolchildren here for the day, Audrey stood stock-still and allowed them all to stroke her simultaneously to their hearts' content. We had feared that such an invasion might have frightened her but she obviously felt it was worthwhile giving up a bit of grazing time for such a good cause.

Sybil was boring: she was a good, sturdy lamb, but if Audrey was Helen Mirren's character from Dennis Potter's *Blue Remembered Hills*, Sybil was just her friend. Audrey liked Sybil and was very nice to her but Piggy was more inventive, better company. Piggy and the lambs went wherever they wanted to. The lambs jumped over the garden wall, and short-legged Piggy pushed the gate open.

When Piggy farrowed it really struck us again how completely she regarded herself as human. The seeming fear she had shown towards her piglets was simply because she had never seen any before. She had been wrested from her mother and siblings when only a few minutes old and had learned to behave as her guardian behaved towards her. She was the only pig I ever knew with table manners. All other pigs rush towards the bearer of food and drink and risk spilling everything in their eagerness, but Piggy would wait until the food was safely in the trough before starting to eat, apparently realising that if it were spilled on the floor it would be lost. Although Piggy learned quickly how to mother her eleven piglets, she did not manage, or perhaps even try, to teach them her good manners.

I remember how one of Piggy's granddaughters, Lucy, managed to disabuse us of our belief in one particular old wives' tale. Rachael and Lucy were busy digging up the small orchard and I was idly wondering what I would do if one of them were to fall in the pond. I felt sure that no pig would jump in voluntarily because they would be bound to know they could not swim, a 'fact' I had 'known' since early childhood. I was supposed to be gardening, but all the time I was half seriously contemplating running to fetch a tractor and a rope when, before my eyes, Lucy launched herself gracefully into the pond, swam round it twice, smiling, got out, shook herself and resumed her rooting. I was transfixed.

One day while I was in the kitchen I heard a very loud banging noise at the back door: a fierce, repeated, relentless rapping. As I rushed to open the door I realised that the banging was accompanied by equally strident and persistent baaing. It was Audrey knocking with her foot. When she saw me she baaed even louder and ran down the lawn, stopped, looked at me, ran towards me, calling, ran away agitated, trying Lassie-like to make me follow. We ran down the lawn, jumped the stream, scrambled up the bank and I found myself standing on the edge of the swimming pool where Sybil was swimming round and round, totally unable to do anything but swim. I jumped straight in to rescue her, put my arms round her and realised immediately that the dry lamb I could carry now had a sodden fleece and was much too heavy for me to lift up the steep sides of the half-full pool. Sybil was cold and distressed, and I had no way of knowing how long Audrey had taken to decide to come and find me. I held Sybil as far out of the freezing water as I could, resting her body on my uplifted knee. Audrey and I both started shouting for my father who I knew was the only person near. He was in the workshop, a hundred yards away and welding noisily. Every shout I made Audrey echoed and sometimes we called together. At last he heard and came hotfoot to our aid. Sybil recovered quickly, though I felt cold for the rest of the day.

Once and only once, a wild lady mallard took her eight newborn ducklings into the three-quarters-full swimming pool but they were too small to get out again. Wild as she was and wary of human beings, she stood on the edge of the pool and quacked and quacked. At first we attached no significance to the noise but eventually the desperate tone got through to us and we went to investigate. She stood her ground. I had my camera with me and thought how happy they looked, and how it was April and I had not planned to have a swim. Eventually, I slid reluctantly in, fully clothed. Every time I stretched out my hand to rescue a duckling it dived and swam the full length of the pool under water. By the time I reached the other end it would repeat the procedure. They were becoming exhausted and so was I. My mother and their mother were waiting. My mother fetched my father and brother and the tennis net. We formed a line across the pool with the net wedged in our toes and held in our hands and slowly advanced towards our 'quarry'. As soon as they were cornered it was easy to catch them. One by one we handed them out to my mother who dried and revived them. The mallard waited patiently, tamely, until all her brood had been returned to her and then she walked them to the safety of the pond.

She did not make the same mistake again, nor have any other ducks since.

Later I shall tell the story of the old grey hen and her bodyguards. This touching relationship arises relatively often. Sheep play bodyguard roles too. When Ellen produced twins at only thirteen months old, her two old great aunts stood guard respectfully and kept the rest of the flock at a slight distance.

Difficult calvings – cows are never wrong

Although some cows hide themselves in remote corners of the farm just before calving, many more find ingenious ways of communicating with us to ask for help. Fortunately there are many more who neither hide nor require help. It is usually the difficult calvings that stay in the memory.

Nonee came down from the high pastures where all her friends were grazing the lush summer grass and lay down in a dark corner of the barn. Such unusual behaviour was sufficient to catch our attention and set us thinking. It was in fact two days before she calved and at that stage the outward signs were not yet obvious, but she felt she needed keeping under observation and of course she was right.

Cows are never wrong about such things.

Black Bumble, one of identical twin daughters of old Mrs Bumble, adopted altogether different tactics to alert us. She too broke away from the herd and came as close to the house as fences would permit, then marched up and down past the kitchen window, about twenty yards out into the field. Her fears were justified as the calf was very badly presented: backwards, upside-down and slightly askew.

Hippolyta deserved sympathy for two reasons. She needed help to calve and, more importantly, as it was the first and only time in her long life that she had had to rely on human beings for anything, she had to work out a way of communicating with them. She simply had not been brought up to it. As far back as she or we could remember, all her family had been fiercely independent. They were a tall family with mottled red-and-white faces, quite gentle and caring to each other but 'a different species from people', to coin a phrase. She really did not want to ask for help but in the end she found a way. She hung around, loitered, got in the way, and was unusually friendly and cooperative. This naturally aroused our suspicions. We tried to make it easy for her; we knew her pride was important. After helping her with a tricky but not painful calving we tried to ignore her as before. We kept half a sly eye on her even so, without her knowing, but she never needed us again.

It is not just when cows are preparing to calve that they seek human assistance.

Over the forty-three years that we have operated a single-suckle system, a number of cows have come to 'ask' to be milked. This can be for a variety of reasons. Sometimes 'teenage' calves can become so keen on eating grass that they have no appetite for milk. In spring, when the grass is superabundant, most cows will produce extra milk. These two factors occurring together can lead to the cow needing to be milked. Some of the cows are even clever enough to ask for help if they develop mastitis while others suffer in silence.

Dizzy and her family

In the late summer of 1966 my parents travelled to the Forest of Dean to collect a two-day-old heifer calf that was being given to them by a man from whom they had previously bought five cows. My father named her Discount. Curled up on the back seat of our blue Ford Cortina, Discount travelled home as good as gold, while my parents listened to the test match with difficulty on a transistor radio held up to the side window, reporting a very exciting last-wicket stand between, I think, Snow and Higgs. Discount was to found quite a dynasty. For one thing she had only daughters and they had daughters too and so the members of the Discount clan multiplied. We still have at least twenty of her direct descendants in the herd today.

One such descendant is Dizzy. Dizzy's first calf was Olé, and he became our herd bull, learning the ropes from Jake from the time he was a few weeks old and Jake was getting on a bit. Olé was magnificent to look at and as nice-natured as his mother, but he was too busy being happy to be as sensitive as Jake. Olé and Mr Mini were exactly the same age and they both wanted to grow up to be just like Jake, who was loved by humans and bovines alike. They had no hope of looking like him, for he was jet Welsh Black, while Mr Mini was mid-Lincoln Red and Olé was creamy Charolais, but they copied what they could. They decided, misguidedly, that he ought to teach them how to fight, and so were perpetually pestering him; attacking simultaneously, banging into his legs or dewlap, reminding him of tickling flies. Jake would just amble around, grazing while the pests jostled and skipped and played for hour after hour until finally, in exasperation, Jake would see them off with a half-push or a stare and they would trot away together.

The story of Dizzy's last calf, Dizzy II, is worth a mention. Dizzy II gained the nickname of The Bodyguard when she was about six months old. Her twenty-year-old mother had begun to suffer from arthritis and had also developed cracks in her feet. As a consequence she was less keen on walking the sometimes great distances covered by the herd in a day and she required embrocation. We visited her several times a day to administer ointment to her legs and linseed oil to her feet and to provide interesting food to save her walking more than necessary. Prior to this, Dizzy II had seen us on average only once a day, and even then we had not been paying particular attention to her mother. These new, more frequent visits aroused suspicion and no matter how far away she might be grazing or playing with her friends, each time we drove up and parked next to her mother, young Dizzy left what she was doing and galloped over to check on us. The older Dizzy was delighted with the attention but the younger always hung around, watching us intently until we had driven away, at which point she would resume her former activities.

Something happens here every day . . .

. . . but many of the happenings, inevitably, go unnoticed.

One prerequisite for greatness, perhaps, is suffering. It brings out the best in people who have experienced war, accidents, loss, injury, poverty, hunger, oppression. If something does not kill you it will root out your depths of goodness, strength and endurance. In the bovine world too, suffering can bring out the best in an individual.

One could say that Black Araminta was a heifer just like any other. It would not be true, but it is the case that we did not realise what she was made of until she had to face adversity.

She broke a bone in her leg – fortunately, and it seems strange to use that word, only thirty yards from the house. This made the task of nursing her much easier than it might have been.

After all the other cows had grazed their way into an adjoining field, we noticed her standing on her own. She did not look at all distressed; she just did not move, and as it turned out she did not move for the next six weeks. She ate and drank enthusiastically and she could lie down and stand up but she made no attempt to go forwards or backwards.

Like almost every other animal, Black Araminta responded beautifully to kindness, and the one exception to this that I remember was her own very last calf, Gemima, whom we had to restrain at two months old so that we could treat a cut foot. Gemima never – and I mean never – forgave what she regarded as demeaning impertinence.

Some animals take to being waited on as if it were their due; others are grateful; some are visibly moved and surprised, but once a routine of grooming and spoiling begins, the healing process is assured. We did not know Black Araminta very well before. She had seemed magnificently independent and capable, as she was to be again, but she put her trust in us absolutely for the duration of her convalescence.

Six weeks to the day after the accident she walked ten yards forward, uphill. We did not witness this. We carried her water buckets uphill too. The next day she walked, in the same direction, fifty yards. The buckets followed suit. On the third day, by the time we looked, she had disappeared from view. We found her in a comfortable hollow, in the shade of a pair of sister oaks, several hundred yards away, and she was not alone.

She had calved unaided and had produced a cream-coloured daughter, Gem; she had licked and suckled her. We approached, to congratulate and investigate, and this was when she made it clear that the period of trust and dependence was at an end. 'Thank you for all you did but from now on, help me only if I ask you to,' she said.

We did as we were told. She produced a calf each year and once in a blue moon needed, asked for and received our help: there was a muddy teat that needed washing, a time when there was too much milk for a very small calf and once a stone in the foot. Apart from that, our docile, compliant, uncomplaining patient became a confident prime minister among cattle: never admitting a mistake or performing a U-turn, never allowing any affection, but vaguely gracious when assisted.

We visit all our cattle at least once every day of the year and if any are close to calving we may go many times. Araminta had a minimum of one opportunity each day to attract our attention and also, like most if not all cows in the herd, she knew where we lived and could always come home if necessary. If she wanted anything, she would march over to greet us as soon as we entered the field. We knew from the rulebook that this was not because she wanted to be stroked. It generally took only a few seconds to discern what was awry.

I wonder if perhaps the very friendliest cows, those who always request affectionate attention, might find it harder to communicate a real need. I shall be on the lookout from now on.

Physical communication

Cows use physical movement of their heads to convey a variety of messages. These movements play a vital role in greeting a person or another animal, when they might stretch their heads forwards, muzzle up, I suppose using their sense of smell. I have never been sure just how big a part smell plays in the everyday life of a cow. Certainly they object strongly if a person wears perfume. But the 'fact' I was told as a child, that cows are colour-blind and recognise each other solely by smell is, I can confidently assert, totally untrue. We have noticed that when a calf falls asleep and its mother grazes her way to the other side of the field, sooner or later one or other of them will demand that they be reunited. If the cow makes this decision first, she will scan the surrounding area, note the various groups of animals and always start walking towards calves of the same colour as her own even if the final choice may be made by smell or some other means. Equally aware, a calf will walk towards cows of the right colour.

A lick of affection together with an eye-to-eye enquiring look might accompany a greeting. A cow such as Gemima will greet all humans with an angry shake of the head and if anyone gets too close they will be physically knocked. This is a warning: it is never followed by an attack, though such warnings, if ignored, can become more insistent.

Heads are used by all members of the herd to greet, recognise and accept into the fold each new calf. Just a simple, quick stare at very close quarters indicates that the new arrival has been logged in the attendance register.

Heads repel unwanted attention but they are also the vehicle for conveying love and concern. Reciprocal grooming is an important activity and the ways in which different cows ask other herd members for such attentions are fascinating.

Notes on grooming

Grooming is a big subject and a very important one. There is only one cow on the farm at the moment who does not like to be groomed by us (and it is not Gemima); all the others, even the grumpy ones, are very appreciative. Our animals keep themselves very clean and if for any reason they are unable to maintain that state they become dejected.

When July Bonnet calved recently– her ninth or tenth calf (I cannot quite remember without the record book beside me) – she contracted a womb infection and felt quite ill for several days. During this time she had no energy for grooming herself and to her absolute satisfaction I took on the task. July is a very big cow and I did not realise the scale of the job until I began, but I could hardly get away with grooming only half of her. Over a week or so our routine developed. Once, she stopped eating hay and I heard a strange noise. It was the noise of contented snoring: July had abandoned herself to the pleasure of being groomed so completely that she had fallen fast asleep.

Cows groom their calves and sometimes, when they are older, calves will groom their mothers in return but quite a lot of grooming goes on every day between animals with no obvious connections. If a cow offers another cow the top of her head, bowed and submissive, she will almost always receive a grooming session in return. (A warning display of aggression with the head held in an almost identical position but with muscles tense and head bowed might appear indistinguishable to a casual onlooker.) All the animals know, without turning round, whether the one walking behind them is friendly or not. I suppose this is not surprising: we all knew in the school playground if friends or enemies were near.

Carrying a brush so that you are always ready to groom a cow can have huge benefits. Apart from the calming effect it can have on a disturbed individual, a situation where some foreign object lodged in a foot might otherwise necessitate a long and painful journey home can often be resolved if the surprise and subsequent pleasure of a grooming affords an opportunity to remove the offending object.

Although some children, like Dizzy II, look after their mothers, some are very selfish. Every day for more than a month now I have been watching a grooming routine take place outside the kitchen window. Laura and her son emerge from their night quarters as soon as the weather tempts them forth and invariably pause within my view for half an hour while mother meticulously grooms son. When she has finished she asks him to groom her in return. He tries hard to refuse and usually succeeds.

Laura lowers her head and pushes him gently; you can almost see the look of disbelief and boredom on his face. He moves away a few inches; she stretches her head towards him. Once in a while he will give her two licks and then stop. She tries every tactic; he remains unmoved. After a time she will bunt his dewlap and push him a bit harder. He bestows one more lick and then asks her to groom him again. At first she refuses but she always gives in. The performance is repeated day after day.

A word about milk

Calves and cows vary as much as people in growth rate, temperament, ingenuity and affection. One interesting factor that might affect growth rate is the marked difference in the taste of milk from different cows. Of course this might have no effect at all but it is nonetheless interesting.

As we are largely self-sufficient in food, we notice the taste of milk probably more than the calves do. They always have the same milk, whereas we taste milk from many cows. It is well known that the milk from different breeds of cow has its own distinctive characteristics of taste and quality. A change in diet will also affect milk taste but there is an inherent difference and we have found that even cows of the same age and breed can produce milk with vastly different tastes and widely variable butterfat contents. In the house we label the milk jugs with the cows' names and we all have our particular preferences.

Calf games

Differences in temperament have become apparent throughout these stories, but I have not yet commented on the behaviour of very young calves. Mersey II, long-awaited daughter of Mersey, last daughter of Meuse, herself one of almost identical twins born in 1969, was a born leader in the realm of calf-game invention. Mr Mint, Dorothy and Seal were all born in December 1994 and little Mersey did not join the gang until 22 February 1995. The gang of three had been relentless in their pursuit of fun and even on bitingly cold days they would do handstands, and leap and skip, and fight their boring, grazing mothers. All of this peep show took place right outside the kitchen window in the newly fenced paddock. Day after day we vowed to buy a video camera but the days passed and the calves gradually slowed down and began to spend more time eating than playing. Then along came Mersey. Her tiny, dull-gold daughter with head held high and an inability to walk at less than a prance decided to show the others what speed really meant. Her prowess and enthusiasm inspired all around her and fresh life was injected into the threesome with new, astonishing games invented by the minute.

As calves get older and genuinely need to spend most of the day eating, they develop a routine for games at dusk. Sometimes the momentum created involves the 'teenagers' and sometimes even the 'old ladies' themselves. This is the time of night when we have seen calves playing tag with a fox, chasing pheasants and organising heats in race-me-to-your-mother-and-back with the eventual winner leading a lap of honour round the perimeter of the field. There have been many pack leaders over the years:

Lochinvar, Isadora, Carpet, Anne, Woolly Bully, Jack, and many taggers-along too.

New moons and full moons punctuate the hurrying months and Gold Belinda has just produced Little Black Belinda but so far has not spoken one word to her. Yet again we humans have the responsibility and delight of providing all the affection and grooming while her mother supplies the milk, which Little Belinda sneaks from behind while her mother is busy eating. Sometimes, when she sees us approaching with grooming brushes in our hands, she is so excited that she starts to bounce on all fours, and gets so carried away that she bounces right past us, then suddenly remembers and rushes back again.

Hens like playing too, and I shall turn to them in a minute. First I must attempt to write about Amelia.

Amelia

Amelia was an unusually delightful calf, more trusting and understanding than we would have thought possible, while her mother was offhand to say the least. From day one, Amelia did everything slowly. She seemed thoughtful. When the gate was opened and all her contemporaries rushed eagerly to the next adventure, Amelia would take her time and emerge, when she felt like it, sometimes when the others were nearly out of sight. She made a mental note of everything, as we were later to find out.

I could write for a thousand pages, listing every detail of Amelia's life, and I still would not have presented an even half-accurate picture of her. She had to cope with the vicissitudes of life, and after giving birth to dead twins her grieving was far more acute than any we had ever witnessed. She needed to be milked each day and during the succeeding year developed a strong friendship with my brother who bent over backwards in an effort to console and divert her. She has always been loving, and though I give her a love only if she looks as if she wants me to, my brother Richard gives her love whether she wants him to or not.

When Richard had officiated at the birth of Nell's beautiful Hereford calf, he went straight over to talk to her calf of the previous year to compensate him for no longer occupying centre stage in his mother's affections. My mother was watching this and noticed that Amelia was watching too. As soon as Richard had finished talking to Nelson my mother told him that Amelia had been watching, somewhat jealously. Richard went straight to apologise to her but as soon as he got near she tossed her head in high dudgeon, turned her back on him and walked off. Richard went after her, grabbed her round the neck and gave her a determined bear-hug then held out his hand for reconciliation. Amelia hesitated for a second then licked his hand and said, slightly huffily, that he was forgiven.

After Amelia had the dead twins, because she had produced so much milk in readiness, she needed to be milked for her own comfort. Richard had to go away from home most days and Amelia was happy for me to bring her *near* home each evening but once we reached the brow of the hill overlooking the farmstead she stopped and insisted on hanging around until Richard returned. She grazed and chatted to her friend but kept an eye out for his little red car. As soon as he drove up the farm road she ambled down and waited in the yard for him. She never mistook any other red car for his.

AMELIA

Patient, loving, knowing Amelia.
Proud, strong, clever, wise,
Able, sure but unconceited.
Endless time for loving your children and us
When we deserve it.
Made noble, brave by pain of loss
Quick to be grateful and happy again.
Straight back, broad muzzle, wide forehead,
Bright eyes, strong legs, well-placed udder,
To all of convention's technical requirements you conform.
Ability to be loved and to be loving,
To make friends, play jokes, take offence,
Solve problems, hog the limelight,

Educate your offspring, be almost too clever sometimes.
Assessing character, distinguishing people,
Not suffering fools at all.
Persevering, being brave, hanging on or
Marching to complain if human intervention's right.
No conceit, one of the herd. Merging in.
Yet standing out, for us, a mile and
Welcoming us into your world.

Hens like playing

Hens like playing. In fact that is all they ever do, apart from eating, which they also seem to do non-stop. They enjoy everything, sing happy little songs and just have fun. From the minute they are let out in the morning they embark on adventures. These include pecking their way to and round and through all the barns, pecking up and round and over all the bales of hay, onto and over the self-feed silage (which term they take literally) ending up in mid- to late afternoon ranged along the top of the wall outside the kitchen window sunbathing, or dustbathing under one of the bushes. They are, however, not over fond of rain.

In the winter the Land Rover is loaded up with hay and the hens try their hardest to cadge a lift. They know they are not supposed to so they peck round the wheels nonchalantly and wait for an opportunity when our backs are turned. One will usually manage to jump in and hide among the hay, and once, when the engine was running and we did not hear the triumphant singing, a hen got away with it and was not discovered until the hay was being unloaded hen-miles away up in the fields. She tumbled out with a bale of hay and the wind buffeted her this way and that. The cows formed a circle of surprise round her but she was not at all intimidated. I picked her up and lifted her into the front of the vehicle, where she stood on the seat and looked round her like a queen on an official drive-about. I resumed the hay distribution and found an egg.

Hens enjoy human company and our hens hate to be left out of apparently interesting conversations. On one occasion we had a large group of French agricultural students here and they had gathered in a ring to learn about crop rotation. The hens felt deliberately excluded and pushed into the centre of the melee. They stretched up to make themselves as tall and noticeable as possible and tried to take part in the conversation the only way they know, by singing loudly.

There's another side to hens

One day I found old Grey Hen on the ground unable to move. The fox had eaten two of her friends and had damaged her leg severely. She was given medical atten-

tion and her leg was bandaged. For three days she ate nothing, however tempting, just taking frequent, small sips of water. On the fourth morning she ate a piece of bread and from then on never looked back, consuming every imaginable delicacy we could think of: raspberries and cream, butter, cheese, wheat, barley, milk, bread and dripping (her favourite), cooked beef, raisins, etc. During the first four days and, as it turned out, for more than a year after, our other two hens exhibited such genuine, loving and altruistic behaviour that we found ourselves marvelling.

They became her devoted bodyguards. When food was offered, they would stand by and watch until she had eaten all she wanted and only then would they eat themselves. They would walk about, pecking and investigating things but every few minutes one or both of them would hurry back to Grey Hen's side to see if she was all right. They comforted her by gently moving their beaks up and down hers. She did not mind them enjoying themselves but became anxious if they went completely out of sight. The two friends, sisters of the same age and much younger than the 'invalid', had not been particularly friendly to the old hen before her trauma. She was truly the *éminence grise* and perhaps they had been kept in line by her to a certain extent. But as soon as she was no longer able to fend for herself they changed utterly.

During all this time she did not move. She sat in a nest of hay in the garden all day and in a different nest in their pen at night. She learned how to ask to be carried to different locations in the garden by craning her neck in the direction her friends had gone, and by looking at us and 'speaking' in a quite unusual and unmistakable voice.

When the day came for the bandage to be removed we had quite a shock: her foot came off in my hand. Grey Hen, however, seemed greatly relieved. The sterile environment inside the bandage had allowed her stump to heal perfectly and as soon as she was replaced in her daybed she promptly walked off. I exaggerate here; she limped off, but so happy to have lost the dead weight on the end of her leg. She used her wings for balance and moved around at will, though still asking to be carried up steps and, to begin with, across the roadway to bed. After a few weeks the end of her stump became sufficiently hardened to make her confident of crossing the concrete roadway herself. Sometimes she would fairly dash across before we could dream of intervening and sometimes she would stay on the soft lawn and 'require' one of us to carry her. She knew whom she could trust.

After years and years of keeping hens we finally had occasion to get involved in their everyday lives. Of all farm animals, hens are usually, if given freedom and access to a wide variety of food and lots of pure water, the most independent creatures and happy to be so. However these three learned how to 'use' us, to our delight and their benefit.

The accident had happened in the spring and once Grey Hen was mobile again she had the summer to look forward to. She hated to miss out on anything but whenever it rained she hid under the garden hammock and if the rain was heavy we brought her in the house. Her friends knew where she was and carried on as usual but when bedtime approached they simply would not go across the roadway to bed without her.

Grey Hen soon accepted the situation: rain equals incarceration. I think she actually enjoyed the comfort. By now she was eating and drinking like a horse and apart from the noisy activity of pecking wheat from a container she was always silent until she decided it was time to go home. If one of us was close at hand she would need only to start manoeuvring towards the door to make her wishes clear. If, however, we were not handy she would resort to whatever tactics she thought necessary to attract our attention.

She would try her special speaking voice first. (She had given up singing.) Once, when that failed, she sidled over to the metal saucepan drawer under the cooker and banged more and more loudly with her beak until we heard. We never kept her waiting again.

Grey Hen enjoyed everything. She grazed the grass on

the lawn voraciously; she investigated or dozed or weeded the rose bed; her friends scratched the earth with their feet and she ate whatever they turned up. Even when they had forgotten to be so polite about food allocation she could always beat them for speed with eye and beak, even if they could still run.

For twenty months Grey Hen thrived but then came the day we had all been waiting for. She ate a modest breakfast but showed no interest in food at lunchtime. She seemed off balance and hung her head. She died with her two close friends and two newer friends whom we had bought a few months after her injury by her side.

The question of whether a hen can grieve had never previously occurred to us. The answer is 'yes'. We wondered whether the two bodyguards would miss their old friend but as we were to see during the succeeding days and weeks, all four were affected.

All the hens had appeared to be happy during the summer but we now know that they had restricted their activities to suit Grey Hen's capabilities. They had turned into couch potatoes.

For the first few days after her death the four hens deliberately congregated in 'her' erstwhile corner of the pen every night. After about a week they did some spring-cleaning and we found to our surprise that the nest and the sack under it had been cleared away and the whole corner turned over and tidied. They were all subdued and shied away from human contact for some time. They also ate a great deal less.

Very gradually they began to resume active life and became more adventurous by the day. We would spot them in unaccustomed locations: down by the pond, in the yard with the cows, down behind the pig pens. Each day they roamed further afield, although we had felt certain that they had been happily suited to the confines of the garden. After three weeks they started laying eggs, resumed their friendly-to-human dispositions and insisted on staying up very late, talking and eating and playing in their pen for at least four hours longer than when the old girl had been one of their number.

Amelia again

Amelia is an important and significant animal who deserves another mention.

It was she who, when very young, persuaded or enabled a hardened-in-his-ways farm worker to like and enjoy the company of cows – finally, and to his immense delight.

He had been taught to talk gruffly to cows, hurry them, let them know who was boss; he had in fact been taught to be afraid of them while never admitting it. He came here for the first time when he was sixty-two to help with a particular job but asked if we had any more part-time work. He was a skilled gardener, so part-time days stretched into mutually beneficial years.

One day when a group of cows and calves (the class of '89) was meandering out for the day, he came to 'help', in a hurrying, anxious, domineering way. I told him he could leave Amelia to find her own way to the field because she liked to investigate absolutely everything en route: stones, bushes, rabbits, people, cars, hens, flowers. Very gradually, his astonishment began to subside and when I added that he could give her a stroke if he liked, he did not do so immediately but went back later. She was only two months old at the time but their friendship had begun and over the next few years he grew to appreciate the members of the whole herd as the diverse and rewarding individuals they are.

In the interest of complete historical accuracy, I have to mention one more Amelia-related incident.

One winter's day when she was eleven, I found her alone with sixteen large round bales of hay. It was more than likely that the gate had been left open and that she was not therefore guilty of breaking and entering, but I nevertheless needed to remove her before she flattened them all (flattening big bales being a favourite game of every bovine if offered the opportunity). Perhaps she was extremely reluctant to move because she had only just arrived, but her instinct to be obedient when faced with a polite request became momentarily clouded and I could see for about a minute that she was in a dilemma. I feel quite certain that she contemplated removing the obstacle to happiness, i.e. me, by pinning me against a wall with her head and perhaps squashing some sense into me. She is a very big and powerful cow and we each stood our ground, looking at each other; there was no escape route and I was frightened. I growled at her to obey and I fancy I could see the thought processes at work. She considered her options and decided against grievous bodily harm. The angry expression that had taken hold of her face relaxed and she turned and walked out of the barn. I immediately furnished her with an armful of the desired hay and she was completely her old self again.

A brief note on birds

All birds are happy, clever creatures. In my experience they all seem to learn by their mistakes and most never seem to make any. Birds always know what the weather is going to do long before anyone else, particularly the Meteorological Office.

From very close quarters we have been able to watch their learning process – or is it part of evolutionary adaptation? Our grandstand view is allowed by the fact that in winter we fill smallish round, plastic cups with breadcrumbs and molten fat which, when set hard, we invert on thin bean canes.

The blue tits learn how to feed from them within minutes, hanging upside-down with ease. Word fairly soon spreads among the rest of the tit population and the cups are visited by marsh, willow, coal and great tits, all of whom devise satisfactory feeding methods within a relatively short time. The chaffinches, on the other hand, begin by being quite cross. They shout at the cups and seem to expect a solution to present itself. After a while they decide to launch themselves in a sidelong direction from the nearest convenient cotoneaster bough and, if lucky, they grab a small beakful. After one or two weeks they hit on the idea of imitating humming birds and approach the cups with beaks outstretched and wings beating desperately fast. This technique is eventually perfected and the 'new' species becomes a regular and successful feeder.

The robin is next to pit his wits and the whole business gives him serious concern. He tries again and again to hang on to the cups long enough to feed but with no luck. We do of course provide liberal supplies of more readily available food and the robin, among others, partakes of this, absent-mindedly, while contemplating the problem of the cups, which were put there in the first place only to ensure that at least some food remained safe from the maggot-pies (as Shakespeare calls them in *Macbeth*), jackdaws and jays. The best the robin could ever manage was to screw himself into a tight ball on the ground directly under the cup and leap vertically, snatching a small amount of food and immediately giving it to his not-quite-so-clever friend who was waiting down below. These spasmodic leaps continued intermittently for a few weeks, then ceased altogether. It was not until May, when the jackdaws were obsessively feeding their young, that one of them learned how to shin up the bean cane and plunder the contents. At about the same time a fox also decided to join in and took the cup away wholesale.

Self-medication

My rather rash statement that 'homeopathy . . . merits a whole chapter in our lives as farmers' is one I shall now modify. While we try to maintain open minds on most subjects and seldom see things as merely black or white, homeopathy has not in fact yet played a very big part in our husbandry techniques.

The adherence to a single method of treating ill animals, whether using so-called conventional modern drugs or homeopathic preparations, herbs, acupuncture, or any of the many other alternative therapies, does not seem appropriate. We try to maintain and promote health and well-being by providing the right living conditions and diet but where something unforeseeable occurs we are willing to try many and varied options and whenever it is necessary we do not hesitate to call the vet and use whatever is prescribed to prevent suffering.

The interesting subject of self-medication in the animal kingdom has until very recently been regarded with scepticism by many scientists but the weight of evidence from observation in the field now makes it impossible to deny that this not unsurprising phenomenon occurs frequently. Because our animals range freely they can and do help themselves to a wide variety of plants.

Undoubtedly taste preferences and curiosity account for some of the nibbling and browsing but I feel certain that when the occasion demands it our animals seek out plants that can help them to recover from illness or injury.

I have mentioned that cows and sheep sometimes eat large quantities of willow, stinging nettles, thistles and ash. The cattle are delighted if ever a tree of any species falls down. Our hens have been seen gorging on digitalis and greater burdock leaves; they may then abstain from both for long periods. Herbal tea made from *Euphrasia officianalis* (eyebright) has greatly improved my own hearing, but there were no detectable signs of improvement for nearly three weeks. It was well worth the wait, and although it is so tempting to opt for the immediate 'cure' or reduction in pain that modern drugs offer, I am hopeful that this slow, helping-the-body-to-help-itself approach might prove more lasting.

My brother Richard wrote in *Country Life*:

Any open-minded practitioner would recognise the fault of modern veterinary medicine in ignoring environmental causes of disease. These are often related to farming systems and the financial pressures which perpetuate them. Vets know there is little use in prescribing fresh air and exercise for a sickly animal locked into a factory farming system. Nevertheless, what makes it particularly hard for the majority to accept homeopathy is the belief that remedies become more potent with greater dilution.

His article was written from a sceptic's perspective of homeopathy, and as this is a farming partnership, we need to find a degree of common ground before contemplating any new ideas. Nevertheless, we are learning gradually and have witnessed some startling recoveries as well as some apparent non-events.

We do, however, applaud the finer points of homeopathic diagnosis. For example, a sweet-natured cow might be prescribed one preparation while a bad-tempered one would be offered something totally unrelated for the same condition. This illustrates graphically that homeopathy recognises and treats animals as individuals. It concerns us greatly that mass medication is so often used in the human and animal kingdoms; vaccines are widespread and no account seems to be taken of individual susceptibilities or the state of immune systems or natural defence mechanisms.

An equally reassuring atmosphere pervades herbalism and several other alternative therapies with which I have come into albeit tenuous contact. The misuse of modern drugs, whether by ignorance or design, has alarming consequences, and it is absolutely certain that intensive factory farming could not continue without an armoury of drugs to keep alive the poor creatures whose quality of life in such systems is non-existent.

Our animals go out of their way to find what they feel they need, and by this I mean actually walking away from the herd in their search. They sometimes pass a more than adequate supply of water on their way to a less handily situated source, which has perhaps a particular mineral content or is a very different temperature.

If given the choice, all farm animals are fussy about the water they drink. Some cows like to drink it as pure as possible, holding their mouths up to a waterfall or pipe, while others will deliberately choose to sip round the 'green mantle of the standing pool', as Poor Tom does in *King Lear*. Sometimes they will wait as much as twelve hours without drinking at all, until they have the opportunity to reach their preferred source.

We are lucky here to have one stream with a very high calcium level; so high that any twigs or acorns that fall in eventually become coated so thickly that they turn into unrecognisable, bone-like objects. There are dew-ponds, pools and a lake and these and the streams give even the most discerning animal a fair choice.

For many years we made a point of offering all our visitors the chance to taste our milk and our water, with some surprising reactions. Quite a large number of people declared they had an allergy to milk but after a short description of how the milk was produced almost all wished to sample it and of those who did not, several asked to take some home to try. We built up a network of friends who could drink our milk but were allergic to any they bought. It seems, from all the observations we have made, that it is likely that allergies are caused not by certain foods but by the way those foods are produced and what treatment the plants and animals from which they derive have received.

There is no getting away from it, appropriate food is the beginning and ending of health. Thomas Sydenham said, 'I had rather undertake the practice of physick with pure air, pure water and good food alone than with all the drugs in the Pharmacopoeia.' *Mrs Beeton's Book of Household Management*, first published in 1861, states, 'By a little care in dieting [the housekeeper] may prevent much outlay in nursing and much money in doctors' bills.' More recently Cindy Engel wrote in *Wild Health* that 'human health is directly reliant on the health of the food we eat

. . . we risk paying for cheap food with our health'.

But somewhere along the way we've lost or ignored this knowledge. Feeding animals is, or should be, instinctively easy: baby blackbirds need worms; lions need meat; sheep and cows need grass. Yet intolerable pressure to cut costs means farmers often trawl international markets for the cheapest and frequently the least appropriate foodstuffs for their animals. If you were to put inappropriate fuel in a car it would perform badly or stop. It seems that the effect of feeding people or animals the wrong food takes longer to discern but the consequences are equally dire and permanent.

More than two-thirds of the farmland in the UK is grassland. Most of this is unsuitable for crop production: keeping cattle and sheep on grassland is the only way to get food from it. We cannot eat grass but they are purpose-built to do just that. At the moment vast areas of arable land are used to grow crops to feed to animals – the least sustainable option. Grassland stores carbon, whereas ploughing releases it into the atmosphere.

Cattle and sheep receive a lot of criticism due to their methane emissions. I am no expert on this, but the one thing I notice that no one ever seems to mention is that when grassland is converted to cropland, the hedgerows get progressively smaller or disappear, often leading to the loss of the large hedgerow trees as well. The role of trees and hedgerows in nature conservation is well known but they are also vital for carbon storage, which at least partly offsets the methane emissions.*

Consumers who actively choose to eat organic high-welfare meat from 100 per cent grass-fed systems, more accurately described as pasture-fed, can thus influence for the better the way in which animals are reared, helping to bring about improvements to their own health as well as to the lives of animals. Such meat is often more expensive but if all the true costs were factored in, it would be less not more expensive and our pastoral landscape would be protected.

My brother Richard works for the Sustainable Food Trust, which is campaigning to increase awareness of the many hidden costs we pay, without realising it, for the way food is produced. More sustainable food production and more stringent animal-welfare systems will become mainstream only once these costs are understood and recognised by society and by governments.

Dorothy and her daughter, Little Dorothy

As a general rule a female bovine would not give birth for the first time before she was twenty-four months old. When Little Dorothy had her first calf she was only fifteen months and still suckling from her own mother.

* Environment Protection Agency, Climate Change Research Programme (CCRP) 2007–2013 Report Series No. 32.

Long before we realised that she was in calf, Little Dorothy had decided that she needed extra food. We would find her in all sorts of unusual places eating hay. She was small and neat and one night she spent in great comfort and solitude underneath a trailer that had hay both on top of and underneath it. The trailer was parked on the roadway that runs through the farm and all the other animals were confined to the fields and barns.

Although we were pleased that she had obviously enjoyed herself enormously, it was very difficult to understand how Little Dorothy had found her way there, and we all accused each other of negligence in having left a gate open. We made sure she was given plenty of time to herself during the day, eating hay ad-lib, and the following night we made doubly sure that the gates were fastened before going to bed. In the morning we found Little Dorothy curled up under the trailer again.

It was more than a fortnight later that I actually saw her escape.

The main gate, which opened onto the road from the field where the Dorothy family lived, was secured by a rope loop that hooked over the shutting post. It seemed like a good fastening, and allowed the gate to be left securely opened or securely closed depending on the situation.

Either Little Dorothy had watched how we fastened it or she had worked it out for herself. By using her nose and patiently wriggling the rope to the top of the post she managed to remove the loop and push the gate open. It always swung to again behind her, which was why no other animals had followed and why everything looked in order each morning.

This was much too good a game to give up and, as we later discovered, she could push her way back in again to see her mother and then go back once more to the trailer.

In May 2002 Little Dorothy gave birth to a tiny black heifer calf. We had been on tenterhooks for days, fearing that she might be too small to manage and wondering whether it might be necessary to perform a caesarean section. In the event everything went well and she required only minimal help.

The next few weeks were an eye-opener for us all.

Old Dorothy had been present during the birth just as Little Dorothy had witnessed her mother giving birth to Luke a few days before that. Old Dorothy was giving her tangible, visible advice and was the most superb grandmother imaginable.

After the initial three or four days Little Dorothy did not have sufficient milk to satisfy her rapidly growing daughter's appetite so we supplemented the diet with bottles of extra milk, taken from another cow with plenty.

There was grass in the fields and both Dorothys ate like caterpillars all day, but it was necessary to bring the new arrival home in order to feed her with milk that had been warmed and put in bottles. The tiny calf understood the routine immediately but her mother (still a teenager by human standards) found it incredibly boring to walk all the way home when what she wanted to do was eat grass out in the fields with her friends.

To begin with Little Dorothy absolutely refused to come in without her mother, so Old Dorothy and Luke came too. We had not considered the possibility of walking the 'baby' home without her mother but it soon became apparent that Very Little Dorothy (as yet unnamed officially) was perfectly happy to walk home with her grandmother. Little Dorothy, therefore, maintained her former lifestyle and although she loved her calf she quite often forgot all about her, leaving her with grandma for increasingly long periods.

Before Very Little Dorothy was three weeks old she had shown herself to be wise beyond her age. She knew why she came home and was perfectly happy to come alone, like a tiny tot being sent shopping with a purse and a list to hand over at the counter of the corner shop. Very Little Dorothy became a fixture in the cow pen at night, preferring to sleep with the older house cows, eat hay at the rack like a grown-up and wander out in the morning to find her mother for her breakfast.

Gradually, Little Dorothy began to take more responsibility for her daughter and as her milk-producing ability improved they spent more and more time together until the bottles of milk were politely but firmly refused.

In the following winter, December 2002, with the whole herd being fed hay every day, the duo thrived in their own special pen. We created an area with an especially low entrance, cordoned off with something a little like a limbo bar, and only the Dorothys understood how to get in and out again. It became their own secret spot where they spent whatever time they needed, in uncompetitive, blissful isolation, rejoining the others at will.

Twenty things you ought to know about cows

1. Cows love each other . . . at least some do.
2. Cows babysit for each other.
3. Cows nurse grudges.
4. Cows invent games.
5. Cows take umbrage.
6. Cows can communicate with people.
7. Cows can solve problems.
8. Cows make friends for life.
9. Cows have food preferences.
10. Cows can be unpredictable.
11. Cows can be good company.
12. Cows can be boring.
13. Cows can be intelligent.
14. Cows love music.
15. Cows can be gentle.
16. Cows can be aggressive.
17. Cows can be dependable.
18. Cows can be forgiving.
19. Cows can be obstinate.
20. Cows can be wise.

Twenty things you ought to know about hens

1. Hens sing when they are happy and enjoy listening to music.
2. Hens chop up food into small pieces for their chicks.
3. Hens 'cluck' constantly to reasssure their chicks.
4. Hens stretch, flap, fly, run, paddle and sunbathe.
5. Hens are fastidious and preen their feathers regularly.
6. Hens dust-bathe as part of their cleaning process.
7. Hens are inquisitive.
8. Hens are playful and can always amuse themselves whatever the weather.
9. Hens are sociable and have many different 'speaking' voices.
10. Hens suffer badly from fear and shock.
11. Hens respond to kindness and attention.
12. Hens love a varied diet . . .
13. . . . and clean, pure water . . .
14. . . . and (preferably) ripe fruit . . .
15. . . . and meat, raw or cooked . . .
16. . . . and some hens like brassica . . .
17. . . . and all like wheat and barley, whole and sprouted.
18. Hens need grit in their diet.
19. Hens make friends . . .
20. . . . and sometimes spurn newcomers.

Twenty things you ought to know about sheep

1. Sheep can be very companionable and amazingly compassionate.
2. Sheep can be highly intelligent.
3. Sheep can be very dim.
4. Sheep always run uphill if they sense danger.
5. Sheep are usually gentle and unaggressive.
6. Most sheep have long, woolly tails to keep them warm.
7. Sheep can live on grass alone but like other things too such as . . .
8. . . . tree leaves and apples.
9. A sheep's thick coat protects it from heat and cold.
10. Sheep can stand very cold weather better than cows, pigs or hens.
11. Some sheep have good powers of concentration and can watch television.
12. Some sheep have butterfly minds and can cause accidents.
13. Sheep seem to prefer running water to still water to drink.
14. Sheep have very long memories.
15. Sheep play almost continually when they are young . . .

16. . . . and sometimes when they are old they have pretend fights.
17. Sheep have several different ways of 'speaking'.
18. Sheep like fresh air and wind.
19. Sheep can be conceited.
20. Sheep can be delightfully affectionate . . . and, of course you cannot pull the wool over their eyes.

Twenty things you ought to know about pigs

1. Pigs take life easy, like comfort and sleep a lot.
2. Pigs like to be waited on.
3. Pigs like being allowed to build nests before farrowing.
4. Pigs cover themselves in wet mud . . .
5. . . . and let it dry and fall off leaving them clean.
6. Pigs are very particular about personal hygiene . . .
7. . . . and always keep their living quarters clean . . .
8. . . . and are the only domestic animal to make a lavatory outside.
9. Pigs 'make' their beds every day.
10. Mother pigs make the beds for their piglets.
11. Pigs need clean water to drink and plenty of water to wash in.
12. Pig's tails are curly if they are happy . . .
13. . . . and straight if they are not.
14. Pigs hate draughts.
15. Pigs can get sun-burned.
16. Pigs are very strong.
17. Pigs are usually gentle and make very nice friends but . . .
18. . . . pigs can be dangerous if threatened or hungry.

19. Pigs need a varied and interesting diet.
20. Pigs always choose the best, most organic food if given the chance.

Kite's Nest Farm

Kite's Nest Farm has been a focus of media attention over the last forty years:

'Don't smile when you read this, because the implications are serious enough, but what the Young family have discovered is this simple fact, cows love each other.'
Adam Nicolson, *Sunday Telegraph*, 1995

'Through long family experience, by intelligence and humanity, Richard Young has shown that organic methods work. Every farm should be like his Kite's Nest. Animals there have space and liberty.'
Jane Grigson, *Observer* Magazine, 1989

'The richest wildflower meadow at Kite's Nest is a remnant of English flora on a scale that is rare indeed . . . they allow livestock to graze it in August only after the flowers have seeded.

The cows are allowed to remain in family groups and their names tend to reflect a family connection.'
HRH The Prince of Wales interviewed in *Highgrove: A Portrait of an Estate* by Charles Clover, 1997

'The Beatles had it right. All you need is love. Rosamund knows every one of her cattle by name, has their monkey puzzle of a family tree in her head and loves them all to bits. Throughout their lives they are subjected to as little stress as possible. They wander as they like from one field to another, choosing whatever grasses and herbs take their fancy, sheltering behind a hedge here, picking a sun trap there. Rosamund can walk up to almost all the animals as they stand in the field . . . and talk to them. As a level-headed observer, I can testify that some come close to talking to her too.

Rosamund swears that one of the more junior cows hung around the house for days with a look on her face which said, "I think I am going to have a difficult calving", before wandering off to do just that.'
Big Farm Weekly, 1989

'The placid natures, glossy coats and sheen of good condition demonstrate more than anything the effectiveness of the traditional management system followed at Kite's Nest.'
Home Farm, 1986

'The beef animals are left to take their own choices as to rearing, feeding and housing. And they make a better job of it than with human interference. In winter they are not forcibly cooped up but are allowed to come and go as

they please . . . Feeding is completely grass-based. The wealth of diet available on the grassland contributes to the health of the animals.'
What's New in Farming, 1986

'All the animals at Kite's Nest have names and very definite personalities.'
Guardian, 1987

'No scientific study of cow psychology could equal Rosamund's account. She has made the term animal sentience come truly alive. We can only be thankful that she has.'
Joyce D'Silva, *Farm Animal Voice*

'The key word at Kite's Nest is contentment. Stress is something to be avoided at all cost and man takes a back seat in nature's affairs . . . Cows boasting sleek coats and sheep whose clean tails look newly brushed wander leisurely around the hills.'
Evesham Journal, 1986

'The cows and calves loomed up out of the darkness for their evening feed of hay, some helping themselves from the back of the Range Rover . . . Rosamund spoke to cows by name as she did the feeding, demonstrating the rapport she has with all the animals at Kite's Nest.'
Gloucestershire Echo, 1988

Bibliography

Anderson, L., *Genetic Engineering, Food and Our Environment*, Green Books, 1999
Balfour, Lady E. B., *The Living Soil*, Universe Books, 1943
Carson, R., *Silent Spring*, Hamish Hamilton, 1963
Cato, *On Agriculture*, Prospect Books, 1998
Cicero, *On Old Age, On Friendship, On Divination*, Harvard University Press, 1923
Engel, Cindy, *Wild Health*, Weidenfeld and Nicolson, 2002
HRH The Prince of Wales and C. Clover, *Highgrove: Portrait of an Estate*, Chapmans, 1993
Harvey, G., *The Killing of the Countryside*, Jonathan Cape, 1997
Harvey, G., *The Forgiveness of Nature: The Story of Grass*, Jonathan Cape, 2001
Lampkin, N., *Organic Farming*, Farming Press, 1990
Mansfield, P., and J. Munro, *Chemical Children*, Century Hutchinson, 1987
Schlosser, E., *Fast Food Nation*, Penguin Books, 2002
Wynne-Tyson, J., *The Extended Circle*, Cardinal, 1990
Young, Rosamund, 'Britain's Largest Nature Reserve?', Soil Association 1991

www.sustainablefoodtrust.org
www.pastureforlife.org